EXPANDING ACCESS TO RESEARCH DATA
Reconciling Risks and Opportunities

Panel on Data Access for Research Purposes

Committee on National Statistics

Division of Behavioral and Social Sciences and Education

NATIONAL RESEARCH COUNCIL
OF THE NATIONAL ACADEMIES

THE NATIONAL ACADEMIES PRESS
Washington, D.C.
www.nap.edu

THE NATIONAL ACADEMIES PRESS 500 Fifth Street, NW Washington, DC 20001

NOTICE: The project that is the subject of this report was approved by the Governing Board of the National Research Council, whose members are drawn from the councils of the National Academy of Sciences, the National Academy of Engineering, and the Institute of Medicine. The members of the committee responsible for the report were chosen for their special competences and with regard for appropriate balance.

This study was supported by Contract No. NO1-OD-4-2149 between the National Academy of Sciences and the National Institute on Aging. Support of the work of the Committee on National Statistics is also provided by a consortium of federal agencies through Grant No. SBR-0112521 from the National Science Foundation. Any opinions, findings, conclusions, or recommendations expressed in this publication are those of the author(s) and do not necessarily reflect the views of the organizations or agencies that provided support for the project.

Library of Congress Cataloging-in-Publication Data

Expanding access to research data : reconciling risks and opportunities.— 1st ed.
    p. cm.
  Includes bibliographical references.
    ISBN 0-309-10012-7 (pbk.) — ISBN 0-309-65340-1 (pdf) 1. Privacy, Right of—United States. 2. Public records—Access control—United States. 3. Freedom of information—United States. 4. Research—Information services. I. National Academies Press (U.S.)
  JK468.S4E96 2005
  323.44'830973—dc22
                              2005030054

Additional copies of this report are available from The National Academies Press, 500 Fifth Street, NW, Lockbox 285, Washington, DC 20055; (800) 624-6242 or (202) 334-3313 (in the Washington metropolitan area); Internet, http://www.nap.edu

Printed in the United States of America

Copyright 2005 by the National Academy of Sciences. All rights reserved.

Suggested citation: National Research Council. (2005). *Expanding Access to Research Data: Reconciling Risks and Opportunities.* Panel on Data Access for Research Purposes, Committee on National Statistics, Division of Behavioral and Social Sciences and Education. Washington, DC: The National Academies Press.

THE NATIONAL ACADEMIES
Advisers to the Nation on Science, Engineering, and Medicine

The **National Academy of Sciences** is a private, nonprofit, self-perpetuating society of distinguished scholars engaged in scientific and engineering research, dedicated to the furtherance of science and technology and to their use for the general welfare. Upon the authority of the charter granted to it by the Congress in 1863, the Academy has a mandate that requires it to advise the federal government on scientific and technical matters. Dr. Ralph J. Cicerone is president of the National Academy of Sciences.

The **National Academy of Engineering** was established in 1964, under the charter of the National Academy of Sciences, as a parallel organization of outstanding engineers. It is autonomous in its administration and in the selection of its members, sharing with the National Academy of Sciences the responsibility for advising the federal government. The National Academy of Engineering also sponsors engineering programs aimed at meeting national needs, encourages education and research, and recognizes the superior achievements of engineers. Dr. Wm. A. Wulf is president of the National Academy of Engineering.

The **Institute of Medicine** was established in 1970 by the National Academy of Sciences to secure the services of eminent members of appropriate professions in the examination of policy matters pertaining to the health of the public. The Institute acts under the responsibility given to the National Academy of Sciences by its congressional charter to be an adviser to the federal government and, upon its own initiative, to identify issues of medical care, research, and education. Dr. Harvey V. Fineberg is president of the Institute of Medicine.

The **National Research Council** was organized by the National Academy of Sciences in 1916 to associate the broad community of science and technology with the Academy's purposes of furthering knowledge and advising the federal government. Functioning in accordance with general policies determined by the Academy, the Council has become the principal operating agency of both the National Academy of Sciences and the National Academy of Engineering in providing services to the government, the public, and the scientific and engineering communities. The Council is administered jointly by both Academies and the Institute of Medicine. Dr. Ralph J. Cicerone and Dr. Wm. A. Wulf are chair and vice chair, respectively, of the National Research Council.

www.national-academies.org

PANEL ON DATA ACCESS FOR RESEARCH PURPOSES

ELEANOR SINGER *(Chair)*, Survey Research Center, Institute for Social Research, University of Michigan
JOHN M. ABOWD, School of Industrial and Labor Relations, Cornell University
JOE S. CECIL, Division of Research, Federal Judicial Center, Washington, DC
GEORGE T. DUNCAN, Heinz School of Public Policy and Management, Carnegie Mellon University
V. JOSEPH HOTZ, Department of Economics, University of California at Los Angeles
MICHAEL HURD, RAND Corporation, Santa Monica, CA
DIANE LAMBERT, Bell Labs, Lucent Technologies, Murray Hill, NJ
KENNETH PREWITT, School of International and Public Affairs, Columbia University
RICHARD ROCKWELL, The Roper Center for Public Opinion Research, University of Connecticut

EUGENIA GROHMAN, *Study Director*
CHRISTOPHER MACKIE, *Study Director (through October 2004)*
MARISA GERSTEIN, *Research Associate*
AGNES GASKIN, *Senior Program Assistant*
ALLISON SHOUP, *Senior Program Assistant (through November 2004)*

COMMITTEE ON NATIONAL STATISTICS
2004-2005

WILLIAM F. EDDY (*Chair*), Department of Statistics, Carnegie Mellon University
KATHARINE ABRAHAM, Department of Economics, University of Maryland, and Joint Program in Survey Methodology
ROBERT BELL, AT&T Research Laboratories, Florham Park, NJ
LAWRENCE D. BROWN, Department of Statistics, Wharton School, University of Pennsylvania
ROBERT M. GROVES, Survey Research Center, Institute for Social Research, University of Michigan, and Joint Program in Survey Methodology
JOHN HALTIWANGER, Department of Economics, University of Maryland
PAUL W. HOLLAND, Educational Testing Service, Princeton, NJ
JOEL L. HOROWITZ, Department of Economics, Northwestern University
DOUGLAS MASSEY, Department of Sociology, Princeton University
VIJAY NAIR, Department of Statistics and Department of Industrial and Operations Engineering, University of Michigan
DARYL PREGIBON, Google, Inc., New York
KENNETH PREWITT, School of International and Public Affairs, Columbia University
LOUISE RYAN, Department of Biostatistics, Harvard University
NORA CATE SCHAEFFER, Department of Sociology, University of Wisconsin–Madison

CONSTANCE F. CITRO, *Director*

Preface

Neither the issue of access to research data nor that of privacy and confidentiality is new; it is not even new to the National Research Council, which has considered one topic or the other, or the two in conjunction, on numerous occasions in the past. Why, then, this reconsideration?

Chapters 1 and 2 offer several answers to this question, based on what has changed in the external environment, especially since the 1993 publication of *Private Lives and Public Policies: Confidentiality and Accessibility of Government Statistics*. But perhaps the most immediate cause is discontent on the part of researchers with the speed and scope of access to the very rich research data that have been collected by federal statistical and research agencies.

Juxtaposed against researchers' demands for increased access are heightened concerns on the part of the agencies and their grantees and contractors about maintaining the confidentiality of their data files. These concerns arise in part from convictions, borne out by research, that perceived risks to privacy and confidentiality reduce survey participation.

In this report, the Panel on Data Access for Research Purposes has tried to reconcile the risks and opportunities arising from increased access by urging a variety of rational solutions: making data available through multiple modes, tailored to the needs of different types of users; undertaking research to improve both the utility and the confidentiality protections of some newer access modes; and measuring the level of research data use as well as the frequency with which confidentiality breaches occur, for example.

Given the panel's relatively narrow charge and limited resources, however, the report does not address certain less obvious, but not necessarily less important, contributors to the problem. One of these is the lack of resources and structural incentives for making data more readily available. At present, outside researchers appear to gain more than agencies do from the prompt and generous release of confidential data, and they stand to lose less than agencies do if such releases lead to documented breaches of confidentiality. Nor does the report examine the reward structures within the statistical agencies, though we suspect that those structures favor data collection over data dissemination. In short, the report proposes solutions that do not require an in-depth look at, and perhaps change of, the motivational structures that undergird the current system of data collection and dissemination.

Nor does the report attempt to decide how much disclosure risk is acceptable in order to achieve the benefits of greater access to research data. Such a decision involves weighing the potential harm posed by disclosure against the benefits potentially foregone. The panel believes that this decision appropriately belongs to the wider community of those potentially affected by it—users, data collectors, and the people who provide the data.

In framing the response to its charge, the panel drew heavily on existing reports and supplemented these reports by commissioning a series of papers on outstanding issues, written by experts and presented at a workshop open to the public. The workshop was held in October 2003 at the National Academies. A summary of the papers presented, together with a list of participants, is included as Appendix A to our report, which we tried to keep quite brief.

Even brief reports, however, make substantial demands on panel members and staff. The panel thanks Christopher Mackie, who served as study director for much of the panel's life, for his critical role in guiding the discussions during its four meetings, for organizing the workshop and writing the summary of the presentations, and for his initial drafting of Chapter 3 of the report. We also thank Eugenia Grohman, associate executive director of the Division of Social and Behavioral Sciences and Education (DBASSE), who was the study director for the final stages of the panel's work and without whose skill, experience, and patience the final report could not have been written. Connie Citro, director of the Committee on National Statistics, provided invaluable guidance and help during the entire process. We are especially indebted to her for incisive contributions to Chapters 2 and 5. We also appreciate the interest and support of Michael Feuer, executive director of DBASSE, and Miron Straf, its deputy director.

Much appreciation is due to the many people who wrote and presented papers at the Workshop on Access to Research Data: Assessing Risks and Opportunities. Their work contributed substantially to our formulation of both the problems and the proposed solutions, as is evident from the citations to their work throughout the report. We are also appreciative of a letter from the Task Force on Confidentiality of the Association of Public Data Users (March 17, 2004) that raised important issues of methods for protecting confidentiality that facilitate data access.

Richard Suzman, associate director of Behavioral and Social Research at the National Institute on Aging, commissioned this study in order to stimulate more creative approaches to the dissemination of research data. We gratefully acknowledge the financial support of the National Institute on Aging.

Finally, I thank the members of the Panel on Data Access for Research Purposes, themselves an extraordinarily knowledgeable, engaged, and vocal group. Together, they represent many disciplines intimately involved in the use and production of research data—economics, political science, statistics, sociology, survey methodology, and law. They represent, as well, differing perspectives, with some being more concerned about expanding access, others about maintaining confidentiality. The panel's discussions reflected these differing experiences and perspectives, and the report tries to balance the competing demands. I appreciate the contributions of all the panel members to this report, but three, in particular, generously contributed to its writing: I thank Joe Cecil, George Duncan, and Kenneth Prewitt for their substantial and indispensable help.

This report has been reviewed in draft form by individuals chosen for their diverse perspectives and technical expertise, in accordance with procedures approved by the Report Review Committee of the National Research Council. The purpose of this independent review is to provide candid and critical comments that will assist the institution in making the published report as sound as possible and to ensure that the report meets institutional standards for objectivity, evidence, and responsiveness to the study charge. The review comments and draft manuscript remain confidential to protect the integrity of the deliberative process.

We thank the following individuals for their participation in the review of this report: Martin David, The Urban Institute; Gerald W. Gates, Policy Office, U.S. Census Bureau; Douglas Massey, Woodrow Wilson School of Public and International Affairs, Princeton University; Trivellore Raghunathan, Department of Biostatistics and Institute for Social Research, University of Michigan; and Avi C. Singh, Methodology Research, Statistics Canada, Ottawa, Ontario.

Although the reviewers listed above provided many constructive comments and suggestions, they were not asked to endorse the conclusions or recommendations nor did they see the final draft of the report before its release. The review of this report was overseen by Richard A. Kulka, Center for Demographic Studies, Duke University. Appointed by the National Research Council, he was responsible for making certain that an independent examination of this report was carried out in accordance with institutional procedures and that all review comments were carefully considered. Responsibility for the final content of this report rests entirely with the authoring panel and the institution.

Eleanor Singer, *Chair*
Panel on Data Access for Research Purposes

Contents

EXECUTIVE SUMMARY		1
1	INTRODUCTION	7
	Scope and Structure of Report, 8	
	Private Lives and Public Policies: A Decade Later, 11	
2	THE CHANGED LANDSCAPE	15
	Increasing Public Concern, 15	
	Society's Increased Need for Data, 17	
	Changes in Information Technology, 20	
	Changes in the Legal Environment, 22	
	Developments in Limiting Data Identifiability, 25	
	Developments in Methods and Procedures for Restricted Data Access, 28	
	Meeting the Challenges, 34	
3	BENEFITS OF ACCESS	36
	Data Collection and Research, 36	
	Data Access and the Scientific Process, 38	
	Microdata for Policy-Relevant Research, 40	
	Longitudinal Microdata, 43	
	Linking Survey and Administrative Data, 45	
	Access and Data Quality, 48	

4	RISKS OF ACCESS: POTENTIAL CONFIDENTIALITY BREACHES AND THEIR CONSEQUENCES	50

Confidentiality Concerns and Nonresponse in
  Censuses and Surveys, 52
Why Confidentiality Breaches Might Occur, 55

5	RECONCILING THE BENEFITS AND RISKS OF EXPANDED DATA ACCESS	63

Documenting Use, 64
Access Through Multiple Modes, 66
Public-Use Data, 69
Facilitating Access to Research Data Centers, 74
Expanding and Improving Remote Access, 77
Licensing Agreements, 78
Informing Respondents of Data Use, 80
Safeguarding Confidentiality: Training, Monitoring, and
  Education, 81

REFERENCES 85

APPENDIXES

A	WORKSHOP SUMMARY	95
B	BIOGRAPHICAL SKETCHES OF PANEL MEMBERS AND STAFF	116

Executive Summary

Policy makers need information about the state of the nation—from the national economy to household use of Medicare—in order to evaluate existing programs and to develop new ones. That information often comes from research based on data collected by statistical agencies or others under a pledge of confidentiality. The most critical data are microdata—data about individual people, households, and businesses and other organizations.

The benefits of providing wider access to microdata for researchers and policy analysts are better informed public policies. The risk of providing increased access to microdata is increased risk of breaching the confidentiality of the data.

Both data collection and research are decentralized activities in the United States. Many federal agencies collect data—from the decennial census to statistics on traffic patterns—and some sponsor data collection through universities and other nongovernment institutions. Although some agencies have in-house staffs of policy analysts and researchers, most researchers are based at universities and other nongovernment institutions. The value of this decentralized system is to ensure a variety of perspectives and approaches to both data collection and research. The challenge is to safeguard the confidentiality of the data while making them available to researchers and analysts in a wide variety of settings. One consequence of the decentralized system is a frequent lack of understanding about how data could and will be used and of planning for those uses.

The charge to the Panel on Data Access for Research Purposes was "to assess competing approaches to promoting exploitation of the research potential of microdata—particularly linked longitudinal microdata—while preserving respondent confidentiality." The panel was asked to consider the tradeoffs between the benefits and risks of data access and to make recommendations about "how microdata should optimally (from a societal standpoint) be made available to researchers."

The panel concludes that no one way is optimal for all data users or all purposes. To meet society's needs for high-quality research and statistics, the nation's statistical and research agencies must provide both unrestricted access to anonymized public-use files and restricted access to detailed, individually identifiable confidential data for researchers under carefully specified conditions.

Research using detailed confidential data is needed not only for well-informed policy making but also to improve the quality of public-use files, which are the most widely used microdata products made available by statistical and other data collection agencies. In turn, wide access to public-use data leads to new analyses and conclusions that must be tested on the more detailed confidential data available only through restricted access.

High-quality public-use files require continuing research into methods of assuring the inferential validity of the data while safeguarding their confidentiality. A great deal of promising work has been done on this topic, but more is clearly needed.

At the same time, the continuing need for restricted access to more detailed microdata means that the conditions for obtaining such access need improvement on a continuing basis. The use of licensing agreements, as a mechanism for granting wider access to confidential microdata, should be expanded. Especially important is easier access to research data centers, such as those maintained at universities and other host institutions by the U.S. Census Bureau. Such centers, which several other agencies maintain at their headquarters, are currently the only place where researchers have access to key microdata that provide the level of detail (e.g., small geographic areas) needed for many important analyses. Research to facilitate secure remote access to these data centers is also needed in order to remove the burden on researchers of traveling to a distant site.

We believe that the changes we recommend will result in wider access to high-quality anonymized public-use files as well as to potentially identifiable microdata. But such expanded access requires expanded procedural and legal protections. The panel believes that users, like agencies, should be held accountable for safeguarding the confidentiality of microdata files to which they are granted access. We recommend that statistical agencies set up procedures for monitoring any breaches of confidentiality that may occur, as well as their causes and consequences. We recommend that agencies require auditing of license holders and penalties for violations of the license. We also recommend that agencies institute confidentiality agreements for public-use data files and meaningful penalties for all data users who willfully violate such agreements.

However, laws, enforcement, and penalties are not enough to safeguard the confidentiality of research records. What is needed in addition to the legal sanctions is a system of norms and values concerning the ethical use of such data. Everyone working with confidential research records—interviewers, data entry clerks, statistical analysts, and social and behavioral scientists—requires education and training in these ethical principles and practices. The statistical system of the United States ultimately depends on the willingness of the public to provide the information on which research data are based. To ensure such willingness, there must be scrupulous attention to assuring the informed consent of data providers, as well as continuing research into public attitudes relevant to data collection, privacy, and confidentiality.

The panel's recommendations should be read in the context of the many existing reports that have addressed similar issues of data access and confidentiality protection in the past. In particular, we have drawn heavily on *Private Lives and Public Policies: Confidentiality and Accessibility of Government Statistics*, published in 1993, though we have not attempted to make recommendations in all of the areas considered in that report. Rather, our recommendations focus on the needs highlighted by legal, social, and technological changes during the last decade.

The panel offers four recommendations on basic issues of documentation and access:

- maintenance of bibliographies of research and policy analysis publications by government and nongovernment data collection agencies in order to provide tangible evidence of the benefits of making data widely available for analysis;
- use of a variety of modes for data access, including restricted access to confidential data and unrestricted access to appropriately altered public-use data, in order to meet research needs for high-quality data with different levels of detail and precision;
- research to guide more efficient allocation of resources among different data access modes; and
- greater involvement of users in planning modes of access to agencies' data in order to better accommodate their needs.

The panel offers four recommendations focused specifically on public-use data. The first two are intended to increase access to data, while the third and fourth try to balance increased access with increased safeguards against misuse:

- research on techniques for providing useful, innovative public-use data sets that increase informational utility without increasing disclosure risk;
- a new system of access to public-use microdata through existing and new data archives (following recommendations in *Protecting Participants and Facilitating Social and Behavioral Science Research*), intended to speed researchers' access to such files;
- a warning on all public-use data that the data are provided for statistical purposes only and a requirement that all users attest to having read the warning; and
- restriction of access to public-use data to those who agree to abide by the confidentiality protections governing such data and the institution of meaningful penalties for willful misuse of those data.

Looking more closely at restricted access to confidential data, the panel offers five recommendations on research data centers, remote access, and licensing agreements:

- for the Census Bureau, broadening interpretation of the criteria for access to data, maintaining a continuous cycle for reviewing research proposals, and taking account of prior scientific review of those proposals in order to facilitate and speed researcher access;
- research by statistical and other agencies that sponsor data collection on cost-effective means of providing secure access through remote data access mechanisms, with the aim of increasing the availability of remote access to confidential data;
- the use of licensing agreements by statistical and other agencies (that do not now have them) for access to confidential data, and expanding the data files for which a license may be obtained;
- development of flexible, consistent standards for licensing agreements and implementation procedures by statistical agencies, with the involvement of data users; and
- inclusion of auditing procedures and appropriate legal penalties in licensing agreements, for the willful misuse of confidential data, in order to balance expanded access with appropriate confidentiality safeguards.

People's perceptions of benefits and trust that they will not be harmed as a result of the information they provide are crucial to their cooperation with data collection efforts. The panel offers two recommendations on this topic:

- provision by data collection agencies of basic information about confidentiality and data access to everyone asked to participate in statistical surveys; and
- continuing research on the views of data providers and the public about research benefits and risks.

Because the panel believes that laws and enforcement alone are inadequate for protecting confidential data, it offers four recommendations on training, monitoring, and education to complement legal, administrative, and technical protections. The first two are directed to data collection agencies:

- providing employees with continually updated written guidelines for confidentiality protection of individually identifiable data and training in confidentiality practices and data management; and
- ongoing research into violations of confidentiality protection procedures and breaches of confidentiality that may occur, as well as the causes and consequences of those breaches.

The second two are directed at educational and professional organizations, which are an important source for the development of professional norms and ethical standards:

- training in ethical issues related to research for all those involved in the design, collection, analysis, and distribution of data obtained under pledges of confidentiality; and
- development of strong codes of ethical conduct to protect the confidentiality of personal data and education about those codes.

The panel is confident that, taken together, these recommendations can improve access to and use of data for research and so improve the quality and relevance of those data for social science research and public policy, while providing appropriate protection for the confidentiality of identifiable data.

1

Introduction

The United States at present has the most extensive array of data collection programs undertaken by federal statistical, research, and administrative agencies in its history. Collectively, these data yield a detailed portrait of population groups and of organizations that affect people's lives (employers, educators, health care providers, and others). When made available in the form of microdata, particularly linked, longitudinal microdata, federal data collections provide an unparalleled resource for policy analysis and research on important social issues.[1] The interest in such data is exemplified by a trend toward studies with great richness and detailed information, such as the proposed national children's study on environmental and genetic effects on health and development (see www.nationalchildrensstudy.gov).

Yet this very trend has increased the risk of violating the confidentiality of those who provide the information. Recent innovations in information technology, such as the widespread availability of data about individuals on the Internet, have also increased that risk. In response, many data collection agencies have reduced the amount of detail in publicly available microdata sets, although they have also worked with researchers to develop new methods and arrangements for data access that protect confidentiality and respect privacy.

---

[1]Microdata, in contrast to summary, aggregated data, provide individual-level information. Linked microdata usually refer to individual survey data that are linked to individual information from administrative records or to other survey data.

Privacy has many dimensions. The emphasis in this report is on informational privacy, which encompasses an individual's freedom to choose the extent and circumstances under which personal information will be shared with others, and how it will be used. Confidentiality refers broadly to an obligation not to transmit identifiable information—for an individual or a business—to an unauthorized party. More specifically, this report is concerned with the explicit or implicit promises made to respondents regarding how their data will be used and the extent to which they will be protected against the risk that the data they provide may allow others to identify them (see National Research Council, 1993:22).

The nation needs to use its statistical data, especially properly protected microdata, for credible, detailed analyses of current and proposed government programs and policies in such areas as education, health care, and taxation. These data are also needed for basic research in the social, behavioral, and economic sciences that can advance the quality and scope of policy analyses. Much basic and policy research will be undertaken outside the federal government, in universities and other research centers. Thus, there are questions about how to provide researchers—inside and outside government—access to data that can both inform public policy and protect the privacy of respondents and the confidential nature of the information they provide.

SCOPE AND STRUCTURE OF REPORT

In response to those questions, the Panel on Data Access for Research Purposes undertook a study to understand and propose ways to resolve the tension between the goal of facilitating researchers' access to federal data collections, particularly detailed microdata sets, and that of maintaining confidentiality. The panel was convened by the National Academies' Committee on National Statistics (CNSTAT) at the request of the National Institute on Aging, which supports the collection of microdata and funds research that depends on the availability of those data for analysis. The panel began its work early in 2003, building on earlier efforts by CNSTAT. Those efforts included a major comprehensive review of the issues more than a decade ago, which produced *Private Lives and Public Policies: Confidentiality and Accessibility of Government Statistics* (National Research Council, 1993) and a workshop held in 1999 (National Research Council, 2000). A CNSTAT report from two decades ago on the benefits of sharing research data is also still relevant (National Research Council, 1985).

The panel was given the following specific charge for its work:

> This study will assess competing approaches to promoting exploitation of the research potential of microdata—particularly linked longitu-

dinal microdata—while preserving respondent confidentiality. The ultimate goal is for the panel to make recommendations about how microdata should optimally (from a societal standpoint) be made available to researchers. This will require, among other things, thinking about how to measure the value of the research good made possible by data production and access, as well as the risk (and associated cost) of disclosures. Such measures are needed in order to assess the tradeoff between the benefits derived from increased protection of data versus those derived from fuller data access.

The panel may also focus on (1) technical, legal and statistical ingredients needed to promote arrangements within and between agencies, and also between government and private sector data producers; (2) enforcement of legal protections for data subjects and appropriate penalties for misuse, and how breakdowns in security are detected, assuming they are, and traced to responsible parties.

The panel will also consider the relative advantages associated with various approaches to data protection and form recommendations about (1) alternative, less burdensome systems (e.g., Internet, remote access, etc.) of providing researchers with access to restricted data and (2) cutting-edge statistical techniques for manipulating data in ways that claim to preserve important statistical properties and allow for broader general data release.

In undertaking its work, the panel quickly discovered that the range of issues, as well as developments since the earlier major CNSTAT report, precluded detailed consideration of the optional elements in our charge. We took as our task a broad overview of the basic issues; we did not explore in detail all data that are or might be available and all ways to protect them. For example, we did not consider the resources and structure of data collection agencies, although we recognize that agencies' histories, priorities, stakeholders, and incentives are factors that affect data access. In addition, although we acknowledge efforts in other countries to develop innovative, workable methods for research access to microdata (notably, the work of the Luxembourg Income Study and the German Socio-Economic Panel Study), we limited our study to the United States because of the differences in this country's laws, organizational structures, and public attitudes. And we touch only briefly on the role of nongovernment survey organizations.

Yet the basic responsibilities and techniques for protecting privacy and confidentiality while promoting data access for research are applicable across all kinds of data, including administrative data on individuals and businesses linked to microdata and individuals' biological data collected in surveys. For example, many kinds of biological information—such as blood pressure, weight, or cholesterol—can be released publicly after some alteration, just as data on income or hours of work can, because the information is not unique. In contrast, genetic data (such as a DNA sample) and data from geographic information systems (GIS) are unique, as are Social Security numbers, and pose much more difficult issues of protection (see National Research Council, 2001a).

This report is intended to take stock of the present situation; it should be seen as one in a line of periodic assessments that will be required over time. It does not pretend to have all the answers nor, given constrained resources, to represent a detailed investigation of alternative data access methods and arrangements. It provides a broad view of the issues, noting why imaginative data access methods are required to satisfy the public need for sophisticated policy analysis and basic social science research.

In addition to regular meetings, the panel held a workshop in fall 2003 to obtain a wide range of views on how issues of data access and confidentiality protection have changed over the past decade since *Private Lives and Public Policies* was published (see Appendix A for a summary of the workshop). The rest of the panel's work was carried out through intense discussions and sharing of draft materials at its meetings.

The rest of this chapter and Chapter 2 provide context for the panel's work. We begin in this chapter with a brief overview of *Private Lives and Public Policies* (National Research Council, 1993). Our report draws on the conceptual framework presented there though it does not revisit in detail the issues or recommendations covered in that study (some of which have not yet been implemented). Rather, our focus is on changes in key areas in the past decade, detailed in Chapter 2: increased public concerns about privacy and confidentiality; society's increased need for data for policy analysis and evaluation and, consequently, for basic social science research; changes in information technology; changes in laws and regulations; and developments in methods for providing access to data for research.

Chapter 3 discusses the benefits to society from the research use of data collected by federal agencies, especially from rich microdata for individuals, organizations, and businesses. Chapter 4 discusses the potential costs to data providers in terms of possible breaches of confidentiality. Chapter 5 proposes ways to reconcile the tensions between the benefits and risks with recommendations for improved access to data for research purposes while protecting the data's confidentiality.

Appendix A is a summary of our workshop, for which we commissioned papers on a range of topics relevant to the panel's task: the economics of data confidentiality (Abowd and Lane, 2003); the role of data access in scientific replication (Bailar, 2003); balancing individual rights and societal benefits from data collection and analysis (Barquin and Northouse, 2003); the role of longitudinal microdata in research and policy

(Brown, 2003); the Census Bureau's research data center network (Hildreth, 2003); recent legislation relevant to privacy, confidentiality, and data sharing (McMillen, 2003); protecting confidentiality of research data through legal means (Perritt, 2003); evaluating inferences from synthetic data (Raghunathan, 2003); estimating probabilities of identification for microdata (Reiter, 2003); monitored remote microdata access systems (Rowland, 2003); licensing and enforcement mechanisms for promoting data access and protecting confidentiality (Seastrom, Wright, and Melnicki, 2003), and the historical record of disclosure and risk (Seltzer and Anderson, 2003). In addition, Michael Larsen gave a talk on technical, legal, and organizational barriers to data linking.

PRIVATE LIVES AND PUBLIC POLICIES: A DECADE LATER

Just over 10 years ago, the Panel on Confidentiality and Data Access produced the report, *Private Lives and Public Policies: Confidentiality and Accessibility of Government Statistics* (National Research Council, 1993). Commissioned by the Committee on National Statistics in collaboration with the Social Science Research Council, the report emphasized the inherent tension between protecting the privacy of individuals and obtaining and disseminating accurate, detailed data to inform public policies. Society affirms to individuals the value of assuring their information privacy and confidentiality, but this affirmation must be balanced with the need of the community for data about individuals and organizations. The first two paragraphs (National Research Council, 1993:15) establish the competing forces:

> Private lives are requisite for a free society. To an extent unparalleled in the nation's history, however, private lives are being encroached on by organizations seeking and disseminating information. In their stewardship of data collection and data dissemination, federal statistical agencies have had a long-standing concern for the privacy rights of the data providers, but they now face mounting demands for privacy . . .
>
> In a free society, public policies come through the actions of the people. Those public policies influence individual lives at every stage—financing of prenatal care, state aid to school districts, job training and placement, law enforcement, and determining retirement benefits. Data provided by federal statistical agencies . . . are the factual base needed for informed public discussion about the direction and implementation of those policies. Further, public policies encompass not only government programs but all those activities that influence the general welfare, whether initiated by government, business, labor, or not-for-profit organizations. Thus, the effective functioning of a free society requires broad dissemination of statistical information.

The report's thorough analysis stressed that government data collection operations must reflect both the obligation to supply the information that is needed to inform the country's democratic and free society and the sometimes competing obligation to respect the individual (or organization) who provides the often highly personal responses on which those data are based. Most importantly, the report laid out a series of recommendations for helping to resolve the tension between these two fundamental mandates. Today, many of these recommendations have been implemented and have led to better information practices. Why, then, look to this issue again?

At root, the tension between the concern for the data provider manifest in the phrase, "private lives," and society's need for data, signaled by "public policies," is structural and can never be fully resolved, no matter how enlightened the practices of a statistical or other data collection agency may be. However, changing conditions can increase (or reduce) the level of tension. As brokers between the data provider and the data user, statistical and research agencies need to continually examine changing conditions and attempt to resolve the resulting tension as best they can.

There is no doubt that these early years in the 21st century challenge statistical and other data collection agencies with a sharply increased level of tension between the two mandates, which, in turn, calls for reexamination of information practices. Several key changes since *Private Lives and Public Policies* thus motivated our study:

- There is evidence of increased public concern about personal privacy and distrust of government assurances of confidentiality; such concern is predictive of reduced cooperation with censuses and surveys (see, e.g., Singer, Mathiowetz, and Couper; 1993; Singer, Van Hoewyk, and Neugebauer, 2003; National Research Council, 2004b; Hillygus et al., 2006).
- There is also evidence of considerable public unease over the burgeoning capability of businesses and private organizations (such as credit rating firms) to gather personal information about millions of individuals (see, e.g., Dash and Zeller, 2005). In early 2001, large percentages of Americans expressed concern about on-line credit card theft (87 percent), Internet fraud (80 percent), hacking of government computers (78 percent), and hacking of business computers (76 percent) (Fox and Lewis, 2001:2). More recently, there have been widely publicized instances of unauthorized release of personal records maintained by large data warehouse firms and credit card companies.
- Assessment of complex public policies requires increasingly de-

tailed data, and researchers increasingly have the ability to carry out a wide range of causal analyses.

- Statistical and research agencies have successfully carried out major data collections, including longitudinal surveys, that can answer important policy questions for which aggregate, cross-sectional data would be inadequate. There is an obligation to ensure that this substantial investment of public monies yields factual evidence that can inform debate in public policy areas.
- New kinds of individually identifiable data, such as unique genetic information and increasingly precise geospatial detail, can be collected and disseminated.
- Statistical agencies, which conduct substantial methodological research on data collection and estimation, often do not have either the internal resources or the political mandate to carry out the causal analyses needed for policy formulation and the advancement of scientific knowledge relevant to policy; this capability is more readily found in the research and policy analysis community.
- Advances in information technology have raised both fears of privacy intrusion and expectations about access to information. Two key factors increasing the risk of disclosure are the existence of comprehensive databases with individual identifiers and the development of sophisticated record-linkage and data-mining methodologies, many of which are readily available on the Internet.
- The legal framework that guides the information process has changed in important ways, notably with the enactment of the Confidential Information Protection and Statistical Efficiency Act of 2002 (CIPSEA) and the fact that medical records, including those used for research, are subject to new confidentiality regulations under the Health Insurance Portability and Accountability Act of 1996 (HIPAA). At the same time, the USA Patriot[2] Act of 2001 overturned the protection previously accorded education records gathered and maintained by the National Center for Education Statistics (see National Research Council, 2005:35).
- New techniques have been developed for producing restricted data products that can be made publicly available because the data have been altered to minimize the risk of individual identification. Techniques have also been developed for analyzing the effects of different alteration methods on disclosure risk and data utility.
- Agencies and researchers now have some experience with restricted data access procedures put in place during the last decade that

---

[2]The name is an acronym for Uniting and Strengthening America by Providing Appropriate Tools Required to Intercept and Obstruct Terrorism.

permit authorized researchers to use confidential data that are not publicly available. Those procedures include protected enclaves, commonly known as research data centers; licensing arrangements; and methods for secure, monitored on-line access to data.

The next three chapters explore these issues in more depth.

2

The Changed Landscape

This chapter details significant changes in the past decade that affect researchers' access to government microdata: increasing public concern about issues of privacy and confidentiality; society's increased need for data; changes in information technology; changes in the legal environment; developments in limiting data identifiability; and developments in methods and procedures for restricted access, including research data centers, monitored remote access, and licensing. We end the chapter with a brief comment on the potential for devising new approaches to data access while taking account of these changes.

INCREASING PUBLIC CONCERN

Private Lives and Public Policies noted public concerns about privacy and confidentiality, but did not describe them in any detail. This report, more so than its predecessor, takes account of changing public attitudes about privacy and confidentiality issues as they bear on principles of data collection and data access.

From recent analyses of data on public attitudes (see Chapter 4), two central findings emerge. First, levels of public concern about the intrusiveness of government inquiries and about whether there is or might be unauthorized disclosure of individual data appear to have increased in recent decades. Second, people who are worried about privacy and confidentiality issues are less likely to cooperate with government surveys. Response to the 2000 census strongly confirmed this second finding, as summarized in Chapter 4 and elsewhere (see National Research Council, 2004b; Hillygus et al., 2006). In addition, there is survey evidence that many members of the public do not believe the government's pledge that data will be kept confidential. In one survey, less than half of the public said that the promise of census confidentiality could be trusted. Also, nearly as many Americans agree as disagree that census answers can be used against them (Hillygus et al., 2006).

A vivid expression of public concern about the privacy and confidentiality of government statistics emerged in spring 2000, when talk show hosts, editorial pages, late-night comics, and even political leaders attacked the 2000 census long form on grounds that it was too intrusive. Responding to a public outcry, President George W. Bush, then a presidential candidate, said he understood "why people don't want to give over that information to the government. If I had the long form, I'm not so sure I would do it either" (Prewitt, 2004:1452). The U.S. Senate passed a nonbinding resolution urging that "no American be prosecuted, fined, or in any way harassed by the federal government" for not answering certain questions on the census long form, in effect telling the public it was acceptable to break the law (Prewitt, 2004). Many more census respondents in 2000 than in 1990 answered long-form questions only selectively, leading to unprecedented levels of imputed values for missing responses (National Research Council, 2004b:283-285).

Relevant research does not draw a clear distinction between the effects of privacy concerns and confidentiality concerns on survey response. However, the long-form controversy suggests that it will be useful in future research to determine when respondents are resisting "unwarranted intrusiveness" simply because they do not like particular questions (a privacy concern) and when they are uncooperative because of fears about "unauthorized disclosure" (a confidentiality concern).

The Census Bureau recognized the importance of this distinction when, in its statement of privacy principles for the general public, first developed in 2000, it acknowledged the importance of balancing the need for statistical information with a respect for individual privacy. The Census Bureau now offers a principle, titled "respectful treatment of respondents," under which it offers two promises (for voluntary surveys): "we promise to respect your right to refuse to answer any specific questions or participate in the survey" and "we promise to set reasonable limits on our efforts to obtain completed questionnaires and will restrict the number of follow-up contacts we conduct" (www.census.gov/privacy/files/data_protection/002822.html).

This newly articulated principle on the part of the Census Bureau is indicative of how much has changed in the few years since *Private Lives and Public Policies* was issued. Few people would have suggested in 1993 that the Census Bureau, which has built its reputation on persistence in getting complete answers from nearly 100 percent of its sample respondents, would only a decade later offer a principle that seems to contradict long-established practices.

SOCIETY'S INCREASED NEED FOR DATA

Public policies very often focus on population groups defined in terms of one or more characteristics: low-income families, veterans, Medicare patients, preschool children, software engineers, drug addicts, homeowners, to name a few from a long list. Policy design proceeds on the basis of knowing how many people are in these groups, how they are geographically distributed, and how they differ in characteristics. Other public policies focus on establishments: small businesses, public schools, military bases, banks, hospitals, prisons, insurance firms, and so on—all entities that are subject to statistical measurement and for which detailed information is required in order to inform policy choices.

Complex policy-making requires multivariate causal thinking about policy alternatives, which, in turn, requires complex, multivariate, often longitudinal data. For example, how will changing the age of eligibility for Social Security affect retirement decisions across different occupations and regions of the country? Over what time-frame (if at all) does Head Start close the educational gap between children from disadvantaged families and children from better-off families? At what levels of state and local unemployment will single mothers in welfare-to-work programs find secure jobs, and what will be the consequences for their children?

As the economy grows more complex and the population becomes more diverse, increasingly detailed data and data analyses are required for policies to match well with economic and demographic realities. This is true not only for policy making, but also for policy assessment and evaluation.[1] A nation learns how well policies are working by comparing their intended effects with the actual outcomes. These comparisons draw on statistical information. One example makes this obvious point. Congress recently debated whether undocumented college-age students who have lived in the country for at least 5 years and performed well in secondary schools should be eligible for in-state tuition if they enroll in a public university in their state of residence (American Association of State Colleges and Universities, 2003). On one side of the debate are people

---

[1] For examples, see National Research Council (1997), which assesses data requirements for the analysis of retirement income policies, and National Research Council (2001b), which does the same for welfare reform policies.

who assert that rewarding undocumented people in this fashion will be an incentive for increased illegal immigration. On the other side are people who assert that there will be long-term economic contributions to society if this group attends colleges and universities. Either, neither, or both of these assertions could turn out to be correct. Evaluating these alternative assertions appropriately requires an exercise in data-based policy analysis.

In addition to federal, state, and local government policy uses of statistical information, commentators on American democracy—starting with George Washington and Thomas Jefferson—have repeatedly stressed that an uninformed public is incompatible with preserving democratic principles and practices. Just as statistical data are used by political leaders to design policy, they are used by the electorate to assess how well those policies have worked and thus to rate the effectiveness of the government, especially when data are available on trends. What are often referred to as social and economic indicators play an important role in democratic accountability, as the public gauges the quality of public life by taking note of what is trending up and what is trending down. Crime rates, air quality, access to health care, charitable giving, education levels, homeownership, out-of-wedlock births, voter participation, and unemployment are illustrative of features of our society that are given public visibility in the nation's official statistics. This public visibility strengthens democratic practice.

The private sector is no less in need of detailed information on the American population. Citing only data from the decennial census, a recent National Research Council report (2004b:66) observed:

> Retail establishments and restaurants, banks and other financial institutions, media and advertising firms, insurance companies, utility companies, health care providers, and many other segments of the business world use census long-form-sample data, as do non-profit organizations. An entire industry developed after the 1970 census to repackage census data, use the data to develop lifestyle cluster systems (neighborhood types that correlate with consumer behavior and provide a convenient way of condensing and applying census data), and supply address coding and computer mapping services to make the data more useful.

The most important source of the information used for policy design and evaluation and the other purposes described above are the more than 70 federal agencies that carry out statistical activities of at least $500,000 per year (U.S. Office of Management and Budget, 2004). These agencies include statistical agencies, such as the Bureau of Labor Statistics (BLS), the National Centers for Education and Health Statistics (NCES and NCHS), and the U.S. Census Bureau. They also include research funding and analysis agencies, such as the Agency for Healthcare Quality and Research, the National Institutes on Aging and Child Health and Human Development, and the National Science Foundation (NSF). In fiscal 2004 these agencies were authorized to spend an estimated $4.8 billion for statistical programs to serve the nation's information needs (U.S. Office of Management and Budget, 2004).

Over a 10-year period, the 2000 census alone cost more than $6.5 billion. This seemingly very high cost pales in comparison with the value of the many uses to which census data are put. The Constitution requires that seats in the U.S. House of Representatives be allocated in proportion to population, for which it mandates an enumeration of the population every 10 years. Two other major uses are redistricting and fund allocations. Congressional and state and local legislative districts are drawn on the basis of census counts for small geographic areas. Currently, federal agencies allocate more than $200 billion of federal dollars to states and other areas by formulas that, directly or indirectly, depend on census data (National Research Council, 2004b:Ch. 2). Across the decade, about $2 trillion in federal funds depend on census data, so that the investment in the census represents only 0.0035 percent of the federal funding based on census results. And this calculation does not take into account state and local expenditures, or the huge business investments in marketing, labor practices, and manufacturing and retail location decisions that rely on census data.

Similarly, other statistical programs provide data that serve many purposes and are collected at a fraction of the dollars at stake in the decisions made on the basis of analyses with those data. For example, the National Assessment of Educational Progress (NAEP), conducted by the NCES, has made and continues to make major contributions to analysis and evaluation of the effectiveness of the nation's elementary and secondary education policies at the federal, state, and local levels (see nces.ed.gov/nationsreportcard).

Society's increased needs for data and their intelligent analysis have been matched by a significant expansion of the analytic capacity found in the nation's universities, policy organizations, corporations, and advocacy groups. The government frequently turns to this private-sector-based analytic capacity—especially to university researchers and analysts in specialized private research institutions—to carry out policy analyses and basic research using government-collected data, particularly microdata on individual units. This behavior is recognition that federal research and evaluation agencies are not funded to take full advantage of the public investment in the collection of the original data on their own. In the case of statistical agencies, they may avoid policy-oriented data analysis so as not to impair credibility, which is based in part on the public's perception of their objectivity (National Research Council, 2005).

An effective public-private partnership between data collection agencies and the research community is a critical element in bringing analyses of complex data, particularly microdata, to bear on policy design and assessment. The partnership is strengthened by continuous improvements in data access, both through public-use data sets and through restricted data access modalities (see below and Chapter 5).

CHANGES IN INFORMATION TECHNOLOGY

The information world now functions through extraordinarily complex networks of humans and computers, capturing enormous numbers of records of personal and organizational information, storing them in data warehouses with capacities in petabytes, analyzing them through sophisticated statistical and data mining tools, and disseminating the results through ever more capable communications media. This explosion in the capability of information technology is evident at each stage of the process of data capture, storage, integration, and dissemination by public and private entities (Duncan, 2004).

As recently as 1993, when *Private Lives and Public Policies* was published, easy access to complex computerized databases produced by federal statistical and other data collection agencies was only a design goal. Now the Web provides access to vast arrays of information from a desktop. For example, through www.census.gov, there is ready access to tables and maps of 2000 census data for all geographies to the block level, as well as access to complex microdata sets through extraction and downloading tools. Other statistical and research agencies also offer Web access to detailed aggregate and microdata.[2]

One useful measure of the extent of this enhanced information technology is reduced cost. In each of the four stages of the process of data capture, storage, integration, and dissemination, advances in information technology have pushed costs lower. Although the picture is complicated by demographic and social factors that drive costs up (such as the declining willingness of the public to respond to telephone surveys), in many ways the cost of obtaining data is much less today than it was 10 years ago. Electronic data capture techniques—such as scanning and computer-assisted interviewing, surveillance by video cameras in buildings and on streets, and satellite imaging—have become commonplace and readily available at moderate cost. Similarly, terabytes of data storage can be purchased for little. By scanning, one terabyte of storage can

---

[2]The Web address for access to all statistical agencies is Fedstats.gov.

hold the contents of 2,000 file cabinets of documents. Ten years ago, such storage would have cost $1 million; about 5 years later the cost was less than $800 (Gilheany, 2000). More recently, Hayes (2002:Fig. 4) estimated the cost of a megabyte at a few tenths of a cent, and Rhea and his colleagues (2003) estimated that the cost of a terabyte will be $100 in 2006. Data integration—that is, consolidating information from heterogeneous databases—is no longer a horrendously complex task, but one that is facilitated by data standards (such as XML), the growth of the Web, and fast and inexpensive data transmission capability. Correspondingly, data dissemination through the Web and electronic mail is now free for all practical purposes.

Lowered cost at each stage of the data process provides benefits to researchers. With lowered costs, researchers and policy analysts can enjoy the prospect of being able to work with new data sources, to use historical records that are maintained indefinitely in user-friendly formats, to create rich contextual databases with relevant attributes, and to disseminate their results quickly throughout the world.[3]

Lowered cost also provides opportunities for "data snoopers," by which we mean individuals or organizations that attempt to identify respondents for purposes that range from curiosity to marketing to pinpointing individuals who may have committed a crime or who may constitute a terrorism risk. In contrast, researchers are not interested in individuals as such, but only in the answers to research or policy questions that can be obtained by analyzing aggregations of individuals' attributes.

Yet the data that are most useful to legitimate researchers typically have characteristics that pose substantial risk of disclosure. Some data characteristics that create vulnerability include:

- detailed geographic information;
- repeated data collection from the same subjects;
- outliers, such as people with very high incomes;
- many attribute variables; and
- complete census data rather than a survey of a small sample of the population.

Data with geographic detail, such as census block data, may easily be

---

[3]One such historical source is the Integrated Public Use Microdata Series (IPUMS), which contains individual records from the U.S. censuses from 1850 through 2000 (see Ruggles, 2000). IPUMS is available on-line at the University of Minnesota (www.ipums.umn.edu) with funding from the NSF and the National Institutes of Health.

linked to known characteristics of respondents, unless steps are taken to alter or mask the data. Longitudinal data obtained in panel surveys, which track entities over time, pose substantial disclosure risk—both because identifiers must be retained by the collection agency in order to recontact respondents and because longitudinal data typically include many more attributes than do one-time surveys. In general, data files containing many attribute variables permit easier linkage with known attributes of identified entities. This problem is magnified when social survey data are linked to unique identifiers, such as genetic data. Furthermore, data from a census or near census pose more disclosure risk for some kinds of data snooping than data from a survey having a small sampling fraction: there is little likelihood, for example, that a record from a small sample survey that has some attributes in common with an identified record from an administrative source is actually unique in the population.

The risk of disclosure has been significantly increased because of the ready availability to data snoopers of external databases on the Web (see Sweeney, 2001). These databases identify persons or other entities by name and location and share with statistical data certain attribute variables that may permit matching the anonymized statistical data with identified data (as with marketing databases). Moreover, software for matching is widely available and easily used. Thus, would-be data snoopers now have a treasure trove of potential methods of infringing on the privacy and confidentiality of subjects of statistical surveys by matching information across databases (Winkler, 1998).

CHANGES IN THE LEGAL ENVIRONMENT

For the reasons outlined above, statistical agencies face increased tension as they try to respond to public policy and research needs for data while protecting the confidentiality of the underlying information. In addition, they are expected to maintain both a high quality of the data they produce and their role at the forefront of scientific data collection (see Groves and Lepkowski, 1985). At their disposal they have an array of legal as well as technical solutions. This legal environment has changed in important ways during the last decade.

Recently, federal statistical data have received broad new statutory protections against traditional threats to confidentiality, but they may be increasingly vulnerable to new threats from statutes intended to enhance national security and government accountability (see National Research Council, 2005:35, App. B). In 2002, Subpart A of the Confidential Information Protection and Statistical Efficiency Act (CIPSEA) established minimum standards for protection of information gathered by a federal agency for a statistical research purpose under a promise of confidentiality. Such information may not be disclosed in identifiable form for nonresearch purposes without the consent of the respondent: nonresearch purposes include administrative determinations, law enforcement investigations, and adjudicatory proceedings. CIPSEA thereby provides statutory protection to the many statistical agencies that previously had only custom or other nonstatutory authority to back up pledges of confidentiality. The obligation to protect research data extends beyond federal agency personnel to include those who contract with the agency to provide statistical research services, such as conducting survey interviews or preparing data products. The U.S. Office of Management and Budget (OMB) is developing guidance for the implementation of Subtitle A of CIPSEA, but such guidance has not yet been issued for public comment.

Subpart B of CIPSEA permits identifiable business records to be shared for "statistical" purposes by the U.S. Census Bureau, Bureau of Economic Analysis (BEA), and Bureau of Labor Statistics (BLS), subject to written agreements that specify the nature of the records, the statistical purposes, and the procedures governing access and security. Such data sharing, which fully maintains confidentiality protection, can support significant improvements in the nation's ability to obtain high-quality data on business formation, internationalization of employment, and other critically important issues for economic policy. A key element in the Census Bureau's data is its business register, which is constructed with data from the Internal Revenue Service (IRS). However, without new legislation (to amend Title 26 of the U.S. Code, which governs access to IRS tax data), the business register and associated data cannot be shared with BEA and BLS.

Medical records, including those used for research, are subject to new confidentiality regulations under the Health Insurance Portability and Accountability Act (HIPAA) of 1996 (P.L. 104-191). Researchers who rely on information collected by health care providers must comply with strict requirements governing the use and disclosure of health care information (see National Research Council, 2003b:117-118). Identifiable medical information may be disclosed for research purposes only with the written consent of the person providing the information or in a limited set of circumstances in which an institutional review board determines that the identifiable medical information is essential to the conduct of the research and the disclosure presents minimal risk to the individual. The researchers must protect identifiable information from improper disclosure and destroy the identifiers at the earliest opportunity consistent with the conduct of the research (45 C.F.R. § 164.512).[4]

---

[4]There is anecdotal information that implementation of HIPAA has caused some difficulties for researchers (see Linet, 2003).

At the same time, new challenges to confidentiality of research records have arisen. As noted in Chapter 1, the USA Patriot Act of 2001 (P.L. 107-56) overturned the strict confidentiality protection of education records gathered and maintained by the NCES, a change in protection that was later reflected in corresponding amendments to the statute governing NCES. The USA Patriot Act allows the Attorney General to petition a court for access to identifiable education records, including those from research, for use in the investigation and prosecution of terrorist activities.

Access to federal research information for nonresearch purposes is also permitted by the Shelby Amendment to the Omnibus Consolidated and Emergency Supplemental Appropriations Act for Fiscal 1999 (P.L. 105-277), which requires the OMB to set forth regulations to ensure that all data that are supported by a federal grant to colleges, universities, hospitals and other nonprofit institutions "will be made available to the public through procedures established under the Freedom of Information Act." The resulting OMB guidelines restricted access to data under the Shelby amendment to published or cited research that has been used by the federal government to develop legally binding regulations and rulings, and noted the exemptions to access under the Freedom of Information Act for information that would result in a "clearly unwarranted invasion of personal privacy, such as records that could be used to identify a particular person in a research study" (for a detailed discussion of this issue, see National Research Council, 2003a). CIPSEA strengthens such an interpretation by prohibiting disclosure of confidential information under the Freedom of Information Act. The validity of the OMB guidelines and the effect of the CIPSEA restrictions are the subject of some dispute and have yet to be tested through litigation.

Federal statistical agencies also confront increased scrutiny about the quality of information that they disseminate to the public, even if the data have not been used as part of the regulatory process. The Information Quality Act (also known as the Data Quality Act, P.L. 106-554, § 515, which was enacted in 2000 as an amendment to an appropriation bill), directs OMB to issue guidelines for "ensuring and maximizing the quality, objectivity, utility, and integrity of information disseminated . . . by federal agencies" to the public, and requires all federal agencies to establish administrative procedures to allow affected parties to obtain correction of information disseminated by an agency that does not meet those standards. The resulting OMB guidelines (U.S. Office of Management and Budget, 2005) define "scientific information" to include "any communication or representation of knowledge such as facts or data, in any medium or form." "Dissemination" is defined as "agency initiated or sponsored distribution of information to the public." Taken together, these definitions would extend the regulations to include agency distribution of public-use and restricted use statistical data sets. Agencies must meet higher information quality standards for distribution of "influential scientific information," which is defined as information reasonably expected to "have a clear and substantial impact on important public policies or important private sector decisions." Agencies that disseminate influential scientific information must conduct a peer review prior to dissemination and reveal the data and methods used to generate the scientific information to the extent necessary to facilitate independent reanalysis, while taking into account privacy, confidentiality and intellectual property rights of those who are the focus of the data.[5] The act has raised concern among some researchers that those opposed to certain agency policy initiatives may challenge the findings and quality of research data as a means of impeding agency regulatory activities. In 2003, some 19 agencies received requests for data correction under this act.[6]

Clearly, the courts can have difficulty in balancing individual privacy against a right to public access. For example, in a recent case (*Southern Illinoisan v. Dept. of Publ. Health*, N.E.2d, 2004 WL 1303656 [Ill. App.5 Dist. June 9, 2004]), a newspaper sought release under the state freedom of information act of cancer registry records for people with a rare form of childhood cancer. The state department of public health opposed release of the information, which included zip code of residence, pointing to another state statute that prohibits the public inspection or dissemination of any group of facts that tends to lead to the identity of any person whose condition or treatment information is submitted to the registry. The agency then supported its claim that release would result in inadvertent disclosure of identifiable information by offering expert testimony in which the expert linked most of the records in a test sample to other data and identified the patients. The court dismissed this demonstration, responding that such identification by one expert was not proof that the records could be readily identified, and ordered their release.

DEVELOPMENTS IN LIMITING DATA IDENTIFIABILITY

The confidentiality of individual data may be protected either by restricting researcher access to such data (restricted access) or by various alterations that limit the identifiability of the data and hence permit them to be made publicly available (restricted data). Statistical agencies and

---

[5]National Academies workshop in 2003, "Peer Review Standards for Regulatory Science and Technical Information," also explored this issue; for a transcript, see www7.nationalacademies.org/stl [June 2005].

[6]See www.whitehouse.gov/omb/inforeg/2005_ch/draft_2005_cb_report.pdf [June 2005].

their contractors currently use both methods (see Cohen and Hadden, 2004).

Historically, beginning with the development of public-use summary files and microdata samples from the 1960 and 1970 censuses, restricted data have represented the most widely used method for facilitating researcher access to complex data (see Dunton, 2000; Gaquin, 2000a, 2000b). Restricted data products play an especially important role in providing research access to data because such products are available to all researchers—both inside and outside the government—for critical assessment and alternative analyses. Restricted data may be in the form of microdata files, which contain transformed or imputed attribute values for a sample of individuals (such as age, sex, race, income, occupation, labor force history) or organizational entities. Restricted data may also be in the form of tabular arrays, such as cross-classifications of income and education for geographic areas.

At present, a wide array of summary and microdata sets are available in public-use form. Reflecting the advances in Internet capabilities, access to such data sets is increasingly provided on-line. Examples include:

- the Census Bureau's American FactFinder for geographic area tabulations from the census and American Community Survey (at www.census.gov) (see Hawala, Zayatz, and Rowland, 2004);
- the NSF's on-line tabulation systems for data on science and engineering personnel and resources (www.nsf.gov. statistics.databases.cfm);
- the DataFerrett System of the Census Bureau, which enables researchers to extract and analyze such complex public-use microdata sets as the Census Bureau's Current Population Survey and Survey of Income and Program Participation, and microdata sets from the NCHS, including the National Health Interview Survey and the National Health and Nutrition Examination Survey (dataferrett. census.gov); and
- the Online Data Analysis System of the National Archive of Criminal Justice Data, which uses software developed by the Computer-Assisted Survey Methods Program at the University of California, Berkeley: the archive contains a wide range of federal, state, and local criminal justice data sets and is maintained by the University of Michigan Interuniversity Consortium for Political and Social Research, with funding from the Bureau of Justice Statistics and the National Institute of Justice (www.icpsr.umich.edu/NACJD).

Although many types of restricted data continue to be available, statistical agencies (in response to the increased threats to confidentiality protection noted above) have curtailed the availability of some data that were previously included in public-use files. For example, public-use microdata samples (PUMS files) from the 2000 census were somewhat more restricted in data content than PUMS files from previous censuses, and, because of state laws, the NCHS no longer makes publicly available linked files of microdata from the National Health and Nutrition Examination Survey and mortality records. (Public-use files of business microdata have never been created because of the relative ease of identifying individual firms.)

Restricted data are created through disclosure limitation techniques, which involve either the transformation of the original data, which is called *masking*, or through the use of the original data to guide the generation of synthetic or virtual data through a statistical technique of multiple imputation. Initially, relatively simple data masking techniques, such as top coding income amounts (that is, assigning all income amounts above a certain value to a single category), were used to generate restricted data products. During the last decade the increasing risks of confidentiality breaches have led researchers to develop increasingly sophisticated methodologies for restricted data products (see, e.g., Doyle et al., 2001; Singh, Yu, and Dunteman, 2003).

As developed in Duncan and Pearson (1991), masking may involve coarsening of the data through various forms of data recoding. Attributes may be deleted or combined, or attribute values may be grouped into categories (or bins), such as broad intervals for asset values. Masking may also involve perturbing the data, for example, by adding random noise or stochastically misclassifying certain entities in a table or by swapping selected data among individual records.

The Research Triangle Institute has developed a set of procedures as a sophisticated approach to masking, called macro-agglomeration with substitution, subsampling, and calibration (MASSC) (see, Singh, Yu, and Dunteman, 2003). Substitution refers to perturbation of data fields, and subsampling refers to suppression of individual records. MASSC was originally developed to create public-use files for the National Survey of Drug Use and Health, sponsored by the U.S. Substance Abuse and Mental Health Administration. An important design goal was to protect against breaches of confidentiality by individuals who know someone in the survey—such as a parent who provided consent for a child to participate. MASSC is able to provide measures of disclosure risk and information loss for a particular application.

The development of a methodology for generating synthetic or virtual data is a relatively recent activity (Rubin, 1993). A key objective of the method is to preserve faithful representations of the original data so that inferences from the synthetic data are as consistent as possible with the inferences that would be drawn from the original data. The method is akin to creating multiple samples from the true population. Estimates from any one simulated data set are unlikely to equal those from the observed data, but by combining estimates from several such samples (hence, "multiple imputation"), it is possible to estimate the true value, as well as the amount of variation produced by three sources of error: sampling the collected data, sampling the synthetic units from the population, and generating values for those synthetic units (Reiter, 2003; Raghunathan, Reiter, and Rubin, 2003). The value and usefulness of synthetic data for inferential analysis, though promising, have not yet been fully studied or determined.

In developing restricted data, researchers are paying increasing attention to methods for systematically analyzing the joint impact of various disclosure limitation techniques on disclosure risk and data utility (e.g., Abowd and Woodcock, 2001). A general framework for such an analysis is the risk-utility (R-U) confidentiality map (Duncan, Keller-McNulty, and Stokes, 2003), which incorporates quantified measures of disclosure risk as well as measures of data utility (Duncan and Lambert, 1986, 1989; Lambert, 1993; Reiter, 2003). Nonetheless, more research is clearly needed to assess the relative ability of different masking methods, and of synthetic data, to reduce the risk of disclosure while preserving data utility.

DEVELOPMENTS IN METHODS AND PROCEDURES FOR RESTRICTED ACCESS

Just as important developments have been under way since the early 1990s in methods for producing restricted data products that can preserve both data confidentiality and utility, so, too, has there been substantial expansion in the repertoire of modes for facilitating restricted access to confidential microdata. (Confidential information refers to any identifiable information, regardless of whether direct identifiers, such as name or address, have been removed from the record.) All such restricted access methods are designed to provide the researcher with data not subjected to the perturbations—variable suppression, top and bottom coding, rounding, swapping, random noise, etc.—found in microdata provided in public-use files.

For several decades, major statistical agencies, including BLS, the Science Resources Statistics Division of NSF, and the Census Bureau, have sponsored fellowship programs through the American Statistical Association and the NSF for researchers to work with confidential data at the agency's site.[7] These programs have been invaluable to the agencies in

---

[7] The American Statistical Association also administers research fellowship programs for the NCHS and the BEA.

obtaining significant commitments of researchers' time to working with an agency's key data sets; however, they provide such access to only a handful of researchers each year. The last decade has seen the development of distributed research data center programs, monitored remote access systems, and licensing to encourage many more researchers to work with confidential data.

Research Data Centers

Stimulated by researchers' interests in analyzing detailedwdata on business organizations for which it is very difficult to create useful public-use microdata products, the Census Bureau's Center for Economic Studies worked with researchers to set up two research data centers (RDCs) in 1994. The Census Bureau currently sponsors eight RDCs, with a ninth scheduled to open in late 2005. These RDCs are operated and funded in a variety of ways: one is run by the Census Bureau itself at its Washington-area headquarters; several are run by consortia of research institutions that include universities, not-for-profit organizations, and government agencies; and several are run by a single university with financial assistance from other sponsors, including the Census Bureau (see www.ces.census.gov/ces.php/rdc).[8] The NCHS established an RDC at its headquarters in Hyattsville, Maryland, in 1998 (see www.cdc.gov/nchs/r&d/rdc.htm), and the Agency for Healthcare Research and Quality opened an RDC at its headquarters in Rockville, Maryland, in 2004 for analysis of confidential information from the Medical Expenditure Panel Survey (see meps.ahrq.gov/DataCenter.htm). BLS maintains three separate RDCs at its headquarters in Washington, D.C., for research using confidential data maintained by the Office of Employment and Unemployment, the Office of Compensation and Working Conditions, and the Office of Prices and Living Conditions (www.bls.gov/bls/blsresda.htm). The approval of the Census Bureau is required for access to some BLS data sets for which the Census Bureau is the data collection agent.

The Census Bureau states that the purpose of its RDCs is to increase the utility and quality of Census Bureau data products by providing confidential microdata to qualified researchers under conditions that do not pose unacceptable disclosure risks. The NCHS offers a similar rationale. To access data through an RDC, qualified researchers submit proposals that are reviewed for feasibility, scientific merit, consistency with the agency's mission, and conformity with confidentiality protection proto-

---

[8]An RDC that was started at Carnegie Mellon in 1996 has since closed.

cols. Work is done on a secure site, with secured computers, and under the supervision of agency staff. All outputs are subject to disclosure review. In addition, researchers using the NCHS site sign a confidentiality protocol. To use a Census Bureau site, researchers obtain a special sworn status as a Census Bureau employee. Violation of the terms of that status subjects the researcher to the same legal penalties as Census Bureau employees: for disclosure of confidential data, a fine up to $250,000, imprisonment for up to 5 years, or both.

The rules governing use of the research data centers are more constraining than those encountered in most research settings. Many researchers are willing to accept those constraints because the RDCs provide unique research opportunities that require access to confidential microdata records. At the Census Bureau sites, for example, researchers have access not only to microdata sets on business establishments, but also to demographic microdata sets (including versions of the Current Population Survey and the Survey of Income and Program Participation) with more geographic and socioeconomic detail than is made publicly available and to linkages of population and economic census, survey, and administrative records data (such as the data sets being assembled by the Longitudinal Employer-Household Dynamics Program). At the NCHS site, researchers can merge that agency's data with their own data files.

The development of the RDC concept—particularly the concept underlying the Census Bureau's program of distributing RDCs around the country—is an important step in providing access to microdata sets that pose particularly difficult challenges of data protection. Hildreth (2003:5) claims that the Census Bureau's "RDC network has . . . led to some of the most innovative social science research currently being undertaken." To establish and operate an RDC, however, is costly for a statistical agency and, in the case of a distributed RDC, for the host institution (see Hildreth, 2003). Indeed, the RDC once located at Carnegie Mellon University was closed because the university was unwilling to continue the required level of financial support.

To recover their operating costs, RDCs must attract a sizable clientele. However, the experience to date with the Census Bureau's RDC network is that the long and arduous approval process may be deterring many researchers from applying, particularly graduate students and junior researchers who cannot afford to lose much time before beginning their work. The amount of time varies by project: an economics data project, for example, takes an average of 7 months for approval, not including proposals that require revision and resubmission (Hildreth, 2003:19). Part of the delay for many projects is the necessity to obtain the approval of another agency, such as the Internal Revenue Service (IRS), for the Census

Bureau's business establishment files (which include information from tax returns).

In addition, in response to a 1999 IRS review of the Census Bureau's protocols for confidentiality protection, the Census Bureau specified that the *predominant* purpose of research using data sets that fall under Title 13 (which governs the Census Bureau's operations) must be to benefit Census Bureau programs. This strict interpretation of Title 13, which applies to most of the demographic and economic data provided through the Census Bureau's RDCs, may deter research that would use Title 13 data in important ways but does not meet the strict criteria for approval.

Hildreth (2003) contends that the criteria and time for approval and the direct costs to the researcher associated with use of an RDC have led to their underutilization. He notes the common view among RDC directors and researchers that the use of restricted data sets, specifically the Longitudinal Research Database and other Census Bureau data, has declined. Recognizing these issues, the Census Bureau has recently indicated its intention to consider ways to streamline the RDC process and to explore the addition of data sets from other statistical agencies to its RDC network, which would increase their attractiveness to researchers.[9]

Monitored Remote Access

Monitored remote access to confidential data is currently implemented in four federal statistical agencies:

- the NCES, which permits access to a range of education files containing confidential data using the NCES Data Access System (nces.edu.gov/das);
- the NCHS, which permits access to almost all of the surveys sponsored by NCHS, including geographic and other detail not contained in public-use data products, through remote access to its research data center;[10]
- the U.S. Census Bureau, which permits users to develop their own tabulations from the 2000 census basic records using the Advanced Data Query System (advancedquery.census.gov); and
- the Economic Research Service (ERS) in the U.S. Department of Agriculture, which recently inaugurated a remote system for statistical

---

[9]This information is from comments by Hermann Habermann, deputy director of the U.S. Census Bureau, at a meeting of the Committee on National Statistics, May 6, 2005.

[10]The NCHS system is sometimes referred to as ANDRE—for analytical data research by email.

analysis of microdata from the Agricultural Resource Management Survey (ARMS).[11]

The pioneer of monitored remote access is the Luxembourg Income Study (LIS) in Belgium, which makes microdata from 66 household income surveys available to researchers from 25 participating countries. LIS began in 1983; its software allows users to submit their own programs using standard statistical software. Output is monitored both electronically and manually to protect confidentiality. The LIS software is also used to provide access to the Luxembourg Employment Study, the German Socio-Economic Panel, and EUROSTAT data (see Rowland, 2003). As described by Rowland (2003:4):

> Remote access systems make it possible for users to analyze restricted microdata without visiting an RDC. The systems used for remote access to restricted microdata are monitored automatically and/or manually for disclosure avoidance. They employ automated and manual filters that block certain kinds of queries and results. The files available are usually edited for disclosure avoidance using the same techniques as those used for public use files. They provide more detail to researchers than public use files, but less detail than is usually available in an RDC. The files reside in the [federal statistical agencies] and extracts of microdata and direct access to the records are not permitted.

Unlike the NCES remote access system, the NCHS system allows users to submit SAS (statistical analysis system) programs by e-mail to produce most kinds of output supported by SAS. Output is returned within a few hours of submission (during work hours). Researchers must obtain approval from NCHS for the proposed analysis, sign an affidavit of confidentiality protection, and pay a minimum processing fee of $500 per month, or they can pay $500 per year for selected files that have been developed for repeat and multiple users (for the details of the NCHS policy, see www.cdc.gov/nchs/r&d/rdcfr.htm; Institute of Medicine, 2005). The most used data file is the National Survey of Family Growth, which contains geographic and other detail not available in the public-use format.

The Census Bureau's Advanced Data Query System enables users to develop their own tables from the full 2000 census complete count and

---

[11]ARMS obtains information from farm households on income, assets, and selected crop practices. Access to the remote system requires a memorandum of understanding for research purposes between ERS and the research institution, an approved research project agreement, and a confidentiality agreement with the National Agricultural Statistics Service. The system software monitors data output for confidentiality protection (see arms.ers.usda.gov).

long-form sample records. Tables must be for standard census geographic areas: to protect confidentiality, data cannot be obtained for city blocks or block groups, unlike the prespecified tables that are available from the American FactFinder. Users must register and log in with a user identification and password; there are no processing costs. As of 2003, more than 500 users were registered to use the Advanced Query System; in comparison, the NCHS remote access system had about 45 users in the period 1998-2003 (Rowland, 2003:15,20).

Monitored remote access has the advantage that a researcher does not have to go to an RDC to make use of confidential data and, in the case of the NCES and Census Bureau systems, output is returned quickly. However, output in those systems is limited to tables. The NCHS system provides more output choices, but it has waiting periods to obtain output. With regard to the efficacy of the disclosure review systems, evaluation (see Rowland, 2003) suggests that they protect well against direct disclosure but not against complementary disclosure (that is, disclosure of the information in a table cell by manipulating other cells).

Licensing

Licensing, the third major mode for restricted data access, was first established in 1989 by the NCES. Other agencies and archives that have licensing procedures include the Bureau of Labor Statistics, the Division of Science Resources Statistics of NSF, the Health and Retirement Study data archive (hrsonline.isr.umich.edu/rda), the University of Michigan National Archive of Criminal Justice Data, and the Wisconsin Longitudinal Study (see www.ssc.wisc.edu/wlsresearch).[12]

The license allows researchers to use nonpublic microdata at their own work site, and thus is the most convenient of the three modalities that have emerged over the last decade, although, to date, it is the least used mode. Applicants submit a research plan that includes justification for the use of confidential data, identification of all persons who will have access to the microdata, and a computer security plan. For successful applicants, a license is signed by an official with authority to bind the university, research corporation, or other government agency to the conditions spelled out in the license. Persons with access to the data also sign affidavits of nondisclosure and agree to unannounced inspections to monitor

---

[12]For a list of all the federal agencies that offered licensing agreements for microdata sets as of 2000, see Seastrom (2001:Table 1); for a comparison of licensing arrangements in the United States with those in other countries, see Seastrom, Wright, and Melnicki (2003).

compliance with security procedures. Most license agreements include severe criminal penalties for confidentiality violations.

License agreements are not available in agencies with existing legislation that places more demanding restrictions on confidential data. The Census Bureau, for example, remains constrained by legislation that restricts access to individual records to sworn officers and employees. Licensing conditions vary from agency to agency, as does the duration of the license agreements. Penalties for violating license agreements are uniformly severe, but the procedures for monitoring the performance of licensees and detecting and taking appropriate action against violations are weak (Seastrom, Wright, and Melnicki, 2003). Audits of data protection protocols have found violations due largely to carelessness; they have not found any actual breaches of confidentiality.

MEETING THE CHALLENGES

The panel's report points to a number of serious challenges at the interface of confidentiality and data access. It places a high value on protecting confidentiality. It also takes seriously the responsibility to assure that the nation's robust research and policy analysis infrastructure has sufficient access to microdata so that it can provide intelligent analysis of social and economic conditions and of the effect of policies designed to improve them.

Although it is easy to agree with the Jeffersonian principle that absent an informed public there is no democracy, it is equally easy to agree with the late Senator Moynihan, who famously justified his vote rejecting Robert Bork for the Supreme Court: "I cannot vote for a jurist who simply cannot find in the Constitution a general right to privacy..." But the Jeffersonian public that needs to be informed is the same public that must supply answers to questions sometimes viewed as infringing on privacy and must be assured that answers given are confidential.

The panel finds in history the warrant for asserting that there are ways to move forward without sacrifice to either the value the nation places on privacy and confidentiality or the value it finds in a data-rich democracy. Statistical agencies, working closely with scholars, have for more than 40 years simultaneously improved the technologies that protect confidentiality and the modalities that provide appropriate access to microdata. Even as some methods are applied to decrease disclosure risk, others have been designed to improve access under carefully controlled conditions.

In response to increased public concerns about privacy and confidentiality and developments in information technology and data availability in the past decade, the statistical and research communities responded quickly with new methods for restricted access modes and restricted data products. The challenge at the present time is to evaluate how well the new methods are working, forthrightly assess problematic areas, and determine ways in which alternative methods can be improved. Nothing in the past suggests that increasing access to research data without damage to privacy and confidentiality rights is beyond scientific reach. This report offers recommendations that, if implemented, will continue the past record of simultaneous improvement along both dimensions. Such improvement will require strong partnership between the research community and statistical and research agencies in the design of innovative research on disclosure avoidance techniques and data access modalities and in the implementation of the advances that result from such research.

3

Benefits of Access

The United States, like all modern societies today, depends on complex data to develop legislation, design policies, and evaluate programs. Although aggregate data are widely available from many federal agencies, especially the U.S. Census Bureau, data in that form do not permit in-depth, multivariate analysis of the trends, antecedents, or possible consequences of social and other phenomena of interest. Such analyses require access to microdata, which permit the use of statistical models to study specific questions. As noted in Chapter 2, most of these analyses are done by outside researchers rather than the agencies. Analysts also need access to microdata in order to evaluate data quality, although some of this work, too, is also done by some agencies themselves. Thus, access to microdata by outside researchers is critical to both substantive and methodological work.

This chapter discusses the role of data access in the scientific process, some of the specific ways access to research data have contributed to policy making, and the role of access in addressing the question of data quality. We begin with a brief consideration of individuals as the source of much data and the structure of the federal government as it relates to research and data collection.

DATA COLLECTION AND RESEARCH

Much of the data needed in a modern society comes from individuals. In many cases, people are willing to provide information because it results in direct financial or other benefit to them. In buying a home, for example, the detailed financial information from prospective buyers allows banks to determine what kind of mortgage they will offer, which, in turn, allows prospective buyers to know what they can afford. In this case, as in many others, high-quality information reduces transaction and other costs, which in turn results in lower prices for consumers. Instant access to personal financial data and streamlined credit-reporting systems have not only enhanced the ability of lenders to assess risk quickly, but also increased competition among lending institutions with the result that mortgage rates have been reduced significantly—by some estimates, as much as 2 percentage points—saving American consumers billions of dollars a year (see McCullagh, 2004). In other sectors, mechanisms such as frequent shopper cards and on-line credit applications have reduced the prices of groceries and Internet purchases.

In the market context, people seem generally willing to accept the underlying rationale for surrendering a degree of privacy and confidentiality—and running the risk that their data will be misused—because the financial and other benefits are personal, immediate, and clear. The benefits of data requested by governments are often less recognizable: they accrue to society as a whole, not just the data providers; they may take years to be realized in the form of new laws or programs; and they may be used in indirect and complex ways that are not obvious. The benefits of supplying information to a grocery store to save 50 cents on a can of tuna fish are transparent; the benefits of supplying data to a statistical agency that may contribute to improved research on retirement decisions that could, in turn, improve the functioning of pension or social security systems are not. The lack of transparency in the value of personal data for societal purposes has two consequences: people may be reluctant to support (through taxes) government data collection, and there is evidence that people are increasingly reluctant to respond to government requests for information (see Chapter 4).

The United States has not only a decentralized federal statistical and data collection system (see National Research Council, 2005), but also a decentralized, pluralistic structure for basic and applied social science research and policy analysis (as well as other kinds of research). Most of that research, supported by federal grants and contracts, is carried out at universities, nonprofit research institutions, and for-profit research companies, although some federal agencies conduct considerable in-house research. State and local governments, private foundations, and corporate and individual donations are also sources of social science research support at universities and private research organizations, including advocacy and public interest groups with a variety of policy preferences and perspectives.

Federal statistical and other data collection agencies also carry out some research and analysis, but the largest share of their budgets is allocated to data collection and processing. In-house research is generally limited to descriptive studies, such as analyses of trends and group comparisons, along with significant methodological research to improve data quality and the effectiveness of data collection. Statistical agencies do few studies that have specific policy-related conclusions, although their work often relates to policy questions. One reason underlying this approach is that the agencies must avoid undercutting their credibility as a source of high-quality, objective information. As a matter of principle, substantive analyses by statistical agency staff should be relevant to public understanding and policy issues but "not take positions on policy options or be designed with any particular policy agenda in mind" (National Research Council, 2005:41). Because the scope of research by statistical agencies is often narrowly focused, data access by other researchers is necessary to ensure that alternative methodologies and uses are fully explored to advance social science knowledge and the design and evaluation of public policies. Research access provides opportunities for disparate academic and policy communities to communicate and learn from one another; it also provides valuable information to statistical agencies about their data.

DATA ACCESS AND THE SCIENTIFIC PROCESS

Empirical science includes not only data collection and use, but also data access and sharing. Data access is especially central to the production of policy-relevant social science research in which microdata, which comprise detailed information about individual units (people, households, firms), and particularly longitudinal microdata, which comprise repeated observations on the same units, play an essential role. A large portion of such data is collected in surveys conducted or funded by government agencies. Many of the raw data produced from these surveys contain individual and group identifiers along with sensitive information; the data are typically collected under a promise of confidentiality and are to be used only for research or statistical purposes.

Almost 20 years ago, a study by the Committee on National Statistics described the benefits of data sharing—some of which apply to data access more generally—and its essential role in science (National Research Council, 1985:9-16; see also Sieber, 1991). Data sharing promotes new research and allows for exploration of new questions without necessitating new data collection. Economies of scale are also created. The same datasets can be used for multiple purposes without substantial new investments: data gathered by researchers to answer one set of questions may be useful to others to answer another. Finding new ways to use existing data may also lessen the need for new collection efforts, which, in turn, reduces the burden on respondents. Sharing data can also lead to file linkages and the creation of new, more powerful datasets for examining public policy issues.

When government-funded research is used for decision making, data sharing allows for analysis of problems by investigators with diverse perspectives. Policy disputes related to interpretation are common, and, with wide dissemination of data to researchers, debate can be better informed. In contrast, much of the policy-related research that is commissioned by private interests is never published, so it cannot be corroborated or extended to new work.

When data are shared along with study results, the research community and data collection agencies can improve and hone their own data collection methods and analytic capabilities. Faulty techniques that might not otherwise have been acknowledged as such can be identified, and techniques that are effective can be promoted (we return to this important point below). Researcher access also makes it more likely that additional information about the statistical procedures that underlie the data, which might otherwise not be completely documented by agencies, is archived.

Perhaps most important, data sharing fosters an open research community and reinforces transparent scientific inquiry. Data sharing allows for verification, refutation, or refinement of original results. In this way, data access safeguards the scientific enterprise by ensuring that other scientists can replicate important findings (see Abowd and Lane, 2003). Although replication is not a common scientific activity in the social sciences, philosophers and historians of science (e.g., Kuhn, 1962) agree that it is an important one. "It is by means of wide and complete disclosure, and the skeptical efforts to replicate novel research findings, that scientific communities collectively build bodies of 'reliable knowledge'" David (2001:2). Moreover, there is a considerable amount of informal replication: when an investigator extends previous results, he or she may begin by trying to replicate the first finding. Replication acts as an important disciplinary device for both academic researchers and government statisticians. In addition, as is evident from news stories of research fraud, scientists have sometimes misrepresented the results of research by altering their data or reporting only some observations. Wide access to research data helps ensure that such misrepresentations, when they occur, will be identified by other researchers.

The U.S. National Science Foundation has required that data used in projects supported by its grants be placed in a publicly accessible archive. There are similar requirements at the U.S. National Institute of Justice and the Robert Wood Johnson Foundation. The National Institutes of Health have pursued policies to encourage data sharing for further analysis or replication studies in such areas as DNA sequencing, mapping information, and crystallographic coordinates (Soete and ter Weel, 2003). Some academic journals promote reproducibility of empirical findings for the articles they publish. However, many journals that have a policy of making the underlying data available (e.g., the *American Economic Review*) waive the requirement if any portions of the data are "restricted use," which undermines the value of the policy.

Some important data sets produced by statistical agencies pose particularly difficult challenges of confidentiality protection: they are therefore accessible only in a restricted access mode—a secure research data center, a monitored remote access arrangement, or through a licensing agreement (see Chapter 2). If scientific replication is to be encouraged, application procedures to use confidential data need to be as streamlined as possible so that researchers, including graduate students and junior researchers, are not discouraged from applying by the length of time and amount of resources required for review (see Chapter 5).

MICRODATA FOR POLICY-RELEVANT RESEARCH

The nation faces a range of complex policy issues, including the provision and funding of health care, education standards, retirement income security, and savings and consumption behavior. Equally complex policy issues are posed by such economic changes as increasing globalization of trade and shifts in the relative importance of various industry sectors. In turn, these changes have widespread ramifications for employment opportunities and income security. Addressing policy issues in these areas requires increasingly sophisticated behavioral modeling, which, in turn, requires detailed microdata, particularly longitudinal microdata.

The more detailed the data, the more utility they have for research. For example, the inclusion of geographic details—such as state, county, or city of residence—in microdata sets from large national probability sample surveys would permit modeling disparities in health, economic, and other outcomes that vary significantly across geographic areas. Similarly, the inclusion of contextual variables for cities and neighborhoods in microdata sets from national surveys would permit analyses of many policy-relevant issues. Some RDCs do currently offer access to geographic detail for selected surveys: for example, the National Center for Health Statistics includes a version of the microdata from the National Health Interview Survey with state and county identifiers at its RDC, and researchers can make special arrangements at the University of Michigan to use microdata from the Panel Study of Income Dynamics with contextual neighborhood-level variables derived from census and other data.

Detailed microdata permit in-depth analyses of socioeconomic trends and their antecedents and consequences. Such analyses require multivariate behavioral modeling, which cannot be effectively undertaken with aggregate data. For example, it has become increasingly clear over recent decades that analysis of aggregate statistics does not give policy makers an accurate view of the functioning of the economy (Abowd and Lane, 2003). Indeed, the creative turbulence that is a hallmark of the U.S. economy and a major contributor to its success is not apparent from macrolevel indicators. Analysis of microdata has revealed how flux in labor markets factors into job creation (Haltiwanger, David, and Schuh, 1996) and how widespread reallocation of factors of production (e.g., workers) from one firm to another firm in and across narrowly defined industries is a major contributor to U.S. productivity growth—more important than investment in equipment and structures (Foster, Haltiwanger, and Krizan, 2001).

Detailed microdata are also needed for modeling economic decisions and other kinds of social behavior. Indeed, research is expanding into areas that were relatively untapped even a few years ago. For example, research using attitudinal information in combination with socioeconomic data about individuals and families to model savings behavior—although anticipated 50 years ago by Klein and Goldberger (1955), who used data on consumer sentiment to forecast consumption—has become a vibrant field in recent years with the expanded collection of microdata and their increasing availability to researchers.

A number of applied microdata examples relate to one prominent public policy issue—population aging (see Woodbury et al., 1999). The issue is at the forefront of public attention because of the changing demographic structure of the U.S. population and the budget pressures that face Medicare and Social Security. To develop policies for an aging population in an informed manner, researchers must assess such trends as increasing life expectancy, changing retirement and savings patterns, changes in pension plans, and declines in employer-provided health insurance coverage. Rapidly changing medical technologies and increasing costs of care add further complexity to the analysis. Microdata for individuals and families can be used to simulate outcomes under different possible policies and to estimate costs and benefits associated with various policy options (see National Research Council, 1991, 1997). Data from the Health and Retirement Study (HRS), for example, have been instrumental in answering such questions as how Social Security benefits interact with pensions and savings in household efforts to finance retirement, how social security age eligibility requirements affect retirement rates and timing, and how changes in out-of-pocket medical expenses affect the use of federal programs.[1]

---

[1]For a full bibliography of research using the HRS data, see the website for the survey at the University of Michigan: hrsonline.isr.umich.edu/papers/sho_papers.php? hfyle=bib_all [May 2005].

Another example of microdata needs can be found in research on pollution abatement. In this case, and many others like it, use of aggregate data leads to biased estimates of relationships among variables because different firms in an industry respond to regulations in different ways. Moreover, when aggregated, industry responses are weighted to represent the universe of firms at a given time. As that universe changes as a result of the entry and exit of firms, the assigned weights will no longer be correct and neither will the analyses based on them (see Abowd and Lane, 2004).

Microdata also allow for a much broader range of analyses than do aggregate data: for example, examining relationships among variables for individual classes of firms in an industry or industries (McGuckin, 1995). The expansion of research on the human dimensions of environmental change is another. Researchers increasingly include individual-level contextual variables in their models—the schools respondents attend, the neighborhoods they live in, the firms they work for, and the people with whom they interact. Linking data on people and their environments—including biological and spatial data—is at the very core of this kind of research (Rindfuss, 2002). More generally, the increasing complexity of social and economic activity requires data that can be used to separate out demographic interactions and economic and ecological effects.

A key characteristic of microdata is that they allow the marginal effects of key variables to be isolated, adjusting for other factors. Research into the relationship between household income and Medicare expenditures is an example: interestingly, the results of recent studies have been mixed. Work by Battacharya and Lakdawalla (2003), which uses household-level data from the Medicare Current Beneficiary Survey linked to Medicare claims records, does not find a positive association between income and Medicare use. In contrast, McClellan and Skinner (2004), using insurance claims and data from the census and the Panel Study of Income Dynamics (PSID), find that households in high-income neighborhoods pay more in Medicare taxes but receive more in benefits. It will take more studies, possibly with other microdata sets, to answer the important policy question about redistribution from poor to well-to-do households through the Medicare system.

The microdata-based literature on Medicare shows other interesting relationships. For instance, a small proportion of the elderly population accounts for a very large proportion of Medicare expenditures, and those who account for a high proportion of expenditures in one year are likely to be above-average users in subsequent and preceding years (Garber, MaCurdy, and McClellan, 1998). The importance of the Medicare program, and research about it, is difficult to overstate: total disbursements in 2002 were $265.7 billion, and its costs are growing faster than those of

Social Security. Understanding Medicare program use, and its correlation with income and health, is critical to understanding its current effects and to making projections and policy recommendations for the future.

LONGITUDINAL MICRODATA

For many research applications, it is desirable to analyze not just microdata, but also longitudinal microdata—repeated observations on the same units over time. For example, decisions by individuals and firms that affect retirement behavior and benefits occur over long periods of time, requiring microdata sets that follow people through their working lives (National Research Council, 1997:70-71). Similarly, understanding the cumulative effects of racial or gender discrimination on employment, health, and other outcomes requires longitudinal data on generations of individuals and families (National Research Council, 2004a:Ch. 11).

Longitudinal data contribute to high-quality research in at least two distinct ways. First, such data allow more accurate estimation than is typically possible with a single cross-sectional survey of such information as transitions between states (for example, a household's income falling below the poverty line), durations in a particular state, and changes in variables of interest. Because shorter recall periods tend to result in more accurate reporting of retrospective information, collecting information each year about the past year's activities will produce more accurate data than asking for, say, a 10-year history (Bound, Brown, and Mathiowetz, 2001). Second, longitudinal data allow a researcher to control for the role of unobserved characteristics in explaining variation in outcomes among individuals, so long as the unobserved characteristics are relatively stable for individuals over time (Brown, 2003).

Longitudinal data generally derive from three sources: surveys, administrative records, and policy experiments. Experiments in such areas as welfare reform usually combine data from surveys and administrative records for baseline information around the time of the intervention with follow-up surveys and administrative records at a later time to assess the effects of the intervention (Brown, 2003; National Research Council, 2001b).

Longitudinal data collections are major investments, and their improvement over time is an important goal to which widespread access can contribute. For example, timely researcher access has led to identification of ways to improve economic measures in the HRS, such as the linking of asset values with income from assets (Hurd, Juster, and Smith, 2003). This progress can be contrasted with the lack of progress in the measurement of pension entitlements: those data are complex but, more important, less available. It has taken many years for researchers to understand their weaknesses, in part because few of them have had access to the data and to become familiar with them (see National Research Council, 1997:97-101).

Nearly anything that is done to facilitate the use of longitudinal data for social science research also has the potential to contribute to informed policy discussion. The list of topics to which longitudinal microdata have been applied is a long one: welfare reform, job training, unemployment insurance, preschool programs, retirement, employer-provided health insurance, policies affecting the disabled, K-12 educational reform, occupational safety, and tax policy. However, because of the time it takes to gather and then analyze the data, they are not always available when policy makers want them.

For policy makers, the best-case scenario may appear to be one in which there is a brief policy intervention, the effects of interest are short run, and the data needed for evaluation are contained in routinely collected administrative records. The experiments that preceded the 1996 welfare reform legislation—which replaced Aid to Families with Dependent Children with the Temporary Assistance to Needy Families (TANF) program—are a good example of this happy coincidence. The main effect that the experiments were designed to assess was the extent to which, and how quickly, the prompt provision of job search assistance would move recipients off the welfare rolls; administrative record-keeping for the experimental programs went a long way toward providing the data needed for this assessment. Yet longer term effects of welfare reform, such as the extent to which former welfare recipients hold jobs for 2 or more years, are also of interest, and these kinds of assessments require longitudinal microdata (see National Research Council, 2001b).

Moreover, the early results of experiments can be misleading, so that an adequate period of measurement is needed before effects can be confidently measured. For example, early analysis of the effects of the Seattle and Denver income maintenance experiments indicated that income support leads to marriage dissolution (Groenveld, Tuma, and Hannan, 1980). Subsequent analysis, however, suggested that long-term stability in marriage was enhanced by income support: what had been captured by the data at the beginning of the experiment was a one-time enabling of divorce for women who had no other source of financial support (Cain and Wissoker, 1990).

Longitudinal data can also contribute to policy analyses when a program continues for a long enough time so that evaluations of its early incarnations can be informative in guiding subsequent decisions. Head Start is an example. The earliest cohorts of Head Start participants have reached adulthood, and later cohorts have progressed far enough in school so that the medium-term effects on achievement can be assessed.

Nevertheless, because Head Start was not designed as an experiment using randomly assigned control groups, the program evaluations have been less clear-cut than they might have been if a field experiment had been conducted.[2]

A third situation in which longitudinal data have timely policy relevance is when an ongoing longitudinal data program contains the information needed to address a current policy question. For example, the PSID was originally created in 1968 to study poverty-related problems. It fulfilled that goal well in its early years, and in its more than three decades of existence it has also been the basis for very informative work on duration of spells of welfare receipt (O'Neill et al., 1984; Hoynes and MaCurdy, 1994; Boisjoly, Harris, and Duncan, 1998; Duncan, 2000). The data are now beginning to be used to assess the effects of more recent welfare reforms.

In 1979 the National Longitudinal Survey of Youth (NLSY) was begun, and it and the PSID extended data collection to the children of original PSID respondents. Both data sets have been able to shed light on the consequences for children of poverty, welfare receipt, and maternal employment. Similarly, the HRS, having continued long enough for its initial cohort to reach retirement age, has become the data set of choice for many policy discussions related to retirement.

LINKING SURVEY AND ADMINISTRATIVE DATA

Thirty-five years ago, interagency agreements permitted the linkage of some microdata from the Internal Revenue Service (IRS), the Social Security Administration (SSA), and the Census Bureau's Current Population Survey (CPS). A publicly available 1973 CPS-SSA-IRS exact-match file was the basis for a major dynamic microsimulation model of social welfare policies and retirement income and was also used to analyze the quality of income reporting in the March CPS. A 1978 CPS-SSA exact-match file was the basis for another microsimulation model of retirement income, although that file was not made publicly available. Because linked data present challenges for minimizing the likelihood of re-identifying individuals, concerns about increasing nonresponse rates to government surveys and, subsequently, legislation (e.g., the 1976 Tax Reform Act, P.L.

---

[2]In general, it is easier to mount evaluations with random assignment for experimental programs than for ongoing ones. For Head Start, if program administrators were to define priority classes of applicants and select randomly when they cannot serve everyone in the highest priority class, this difficulty could be overcome. See also the report of an evaluation of the High/Scope Perry Preschool Project, which did use random assignment and found substantial positive effects after an interval of some 35 years (Schweinhart, 2004).

94-455) led agencies to curtail the development of linked microdata for public use. The linkages that were performed (for example, of March CPS files with limited tax return information) were for internal agency use only (see National Research Council, 1991:66-68, 134-135).

Linking survey and administrative microdata can create datasets that facilitate a broad spectrum of research relevant to complex policy questions. For example, by linking Wisconsin income tax records, Social Security earnings and benefits records, and probate records, Menchik and David (1983) determined that prospective Social Security benefits did not have a perceptible effect on lifetime wealth accumulation. Linkages such as those of HRS with Social Security records introduce detail to the data that are particularly constructive for modeling savings incentives, retirement decisions, and other dynamic economic behavior. In this case, the potential research benefits of linking were sufficiently large that the Social Security Administration approved the link with minimal controversy (the linked data are available through special access arrangements at the University of Michigan).

Another example that illustrates the benefits of data linkage is research to investigate the effects of community context on child development and socialization patterns and the effects of the availability of child care on parents' work decisions (Gordon, 1999). By having access to an NLSY file with detailed geographic codes for survey respondents, Gordon was able to add contextual data—such as the availability of child care—for the neighborhoods in which respondents lived. Gordon's application highlights the tradeoff between data precision and disclosure risks. Access to census tract-level geocoding permitted more sensitive construction of community and child care variables central to the study; however, it also increased the identifiability of individual NLSY records.[3]

The new Medicare drug benefit provides another example in which data linkage might contribute significantly to policy-relevant research and modeling. Policy makers would like to know the effects of the legislation: who gains and by how much and how the benefit changes drug consumption, retirement decisions, and other behaviors. To observe and model these responses, individual survey data and linked Medicare data are needed on the same people both before and after the policy change.

Data linkage also has the potential to reduce data collection costs—both direct costs and the cost of respondent burden. Linking existing information from different surveys or a combination of survey and admin-

---

[3] It is important to keep in mind the distinction between identifiability of individual records and the risk of re-identification. The first is a statistical matter; the second assumes, in addition, intruder motivation (see Chapter 4).

istration data, as in the continuing Medical Care Expenditure Panel Survey (MEPS), can streamline the data collection process by reducing the need to duplicate surveys. If survey designers know that links to administrative data can be made, they can limit the length of questionnaires as well. Yet statistical agencies have not made extensive use of linkages of administrative records and survey data in household surveys. They have made much more extensive use of administrative records, such as tax records, for business data collection.

Another benefit from linkages of survey data and administrative records is that they can improve data accuracy and scope by giving researchers access to information that individuals may not be able to recall or estimate accurately. For example, the lifetime earnings data in the Social Security files that are linked to the HRS are virtually impossible for respondents to recall or even find in their own records. Similarly, respondents may not have immediate access to information in medical records or have the technical knowledge to answer some questions. Yet administrative data contain their own errors—such as omissions and duplications—and they may use different concepts and cover different populations than surveys. Thus, although there are many advantages of linked data, care is needed in making linkages and in using the linked data.

In addition to their role in complementing survey data, administrative records can be used to estimate the measurement errors in survey reports, an idea that dates back at least to Ferber (1966) and the work of the Inter-University Consortium for Savings Research. For example, for the Survey of Income and Program Participation (SIPP), a census of federal and state administrative records was taken in 1983-1984 for four states to ascertain the validity of reporting for eight income maintenance programs (Marquis and Moore, 1990). Analysis of survey reports and administrative data for the Food Stamp Program determined that systematic differences in reporting biased the relationships derived from the SIPP sample and that error-prone respondents were more likely to drop out of the SIPP panels.

Another way in which linkage increases data utility is by making it possible to get more research mileage from isolated datasets that would otherwise have limited application. In addition to the above-noted value in reducing data collection redundancies and improving data accuracy in a cost-effective manner, linking information sources may provide increased flexibility to meet future research needs with existing data sets. For example, if SIPP records were linked to administrative tax returns or wage and salary data from state unemployment insurance records, the income data would be more accurate, the cost of the survey would decrease, and the utility of the data for research and policy analysis would increase.

The benefits of linked data extend beyond social science research. Linking records is essential for high-quality studies of effectiveness in many areas of medical research by permitting the analysis of data already available. For example, linking together emergency medical service (EMS) data, hospital records, and death registries allows researchers to follow patients through the pre-hospital, hospital, and post-discharge stages (National Highway Traffic Safety Administration, 2001). Data linkage also facilitates research on infrequent events, such as rare diseases that affect only a small percentage of the population. In such cases, working from general sample data does not provide adequate sample sizes for target groups, and population-based data, which are very expensive to collect, are often required. Linkage can, in some instances, provide a much less costly substitute.

ACCESS AND DATA QUALITY

Researchers' access to and use of the complex data collected by federal statistical agencies are essential to maintain and improve data quality (Abowd and Lane, 2004). The findings derived from such analyses undergo reexamination and reinvigoration when disseminated to the research community. Researchers' use of government data creates an effective feedback loop by revealing data quality and processing problems, as well as new data needs, which can spur statistical agencies to improve their operations and make their data more relevant. The use of data by outside researchers can also verify or improve sampling frames for surveys and censuses. Data access and use by a variety of researchers, for diverse purposes, provides a range of feedback to data collection agencies. Agencies may also be able to generate new data products when they combine existing records with other information from research.

The relationship between data use and data quality is the essential foundation for the common interest of the statistical system and the wider research community in broad and responsible access to data. That relationship is well recognized by such agencies as the Census Bureau. The agency's Center for Economic Studies web page states: "Exposing to the light of research the conceptual and processing assumptions that are embedded in the Census Bureau's micro databases constitutes a core element in the Census Bureau's commitment to quality" (see mission statement at ces.census.gov). This recognition is not new. McGuckin (1992) argued more than a decade ago that coordinated research efforts between in-house and outside researchers offer the best model for ensuring that agencies maximize the benefits from data users. In fact, McGuckin (1992:19) argued that it is a primary responsibility of statistical agencies to facilitate researcher access to confidential microdata files. Such access, by improving the microdata for research and policy analysis, also improves the quality and usefulness of the aggregate statistics on trends and distributions that are the bread and butter of statistical agency output.

Thus, the benefits from research access to complex microdata accrue not only to policy makers, but also, and importantly, to the statistical agencies themselves. There is growing appreciation for the point of view that the largest single improvement that the U.S. statistical system could make is to enhance the capabilities for analysis of statistical data by researchers inside and outside of government, which, in turn, would enable statistical agencies to better understand and improve their data (see Abowd and Lane, 2003).

There are many examples of the synergy that can be created from large numbers of researchers using the same datasets, which allows for corroboration of results and an accumulation of the benefits of knowledge. In a workshop featuring results from Wave 1 of the HRS, for example, much was learned about what had gone right and wrong (*Journal of Human Resources*, 1995). On a practical level, multiple users assure an increased return on the investment in expensive data collection projects.

Although statistical agencies expend substantial resources to ensure that they produce the best possible product, there is no substitute for actual research use of microdata to identify data anomalies. Indeed, there is general recognition of the direct correlation between the quality of a statistical agency's data and its openness to external research. A variety of studies offer evidence that the U.S. statistical system now collects more relevant and higher quality statistics as a result of disclosing both the survey instruments and the data to outside researchers (e.g., Levitan and Gallo, 1990; McGuckin and Nguyen, 1990; Taeuber, 1981; Triplett, 1991; see also Soete and ter Weel, 2003). By increasing access to their data, statistical and other data collection agencies will almost certainly improve both the quality of the data and their usefulness for research and policy analysis. Such improvement will, in turn, increase the value of the investments in collecting, processing, and maintaining the data.

4

Risks of Access: Potential Confidentiality Breaches and Their Consequences

Chapter 3 has argued that to fulfill their function in a democratic society, statistical and research agencies must provide access to the data they collect. Yet, at the same time, they are charged with protecting the data's confidentiality. That charge rests on three underlying considerations: ethical, legal, and pragmatic. The ethical obligation, rooted in the Belmont Report (National Commission for the Protection of Human Subjects of Biomedical and Behavioral Research, 1979), requires agencies to strive for a favorable balance of risks and harms for survey respondents. Legally, they are bound by federal laws to honor the promises of confidentiality they make, with potential civil and criminal penalties if they fail to do so. On a pragmatic level, their ability to collect high-quality data from respondents will be compromised by real or perceived breaches of confidentiality. This chapter elaborates on all three of these assumptions.

A pledge of confidentiality stipulates that publicly available data—whether summary data or microdata and including any data added from administrative records or other surveys—will be anonymized or otherwise masked to ensure that they cannot be used to identify a specific person, household, or organization, either directly or indirectly by statistical inference. Such a pledge also means that more readily identifiable data will be made available for research purposes only through restricted access modalities that impose legal obligations and penalties to minimize the risk that researchers with access to such data might disclose them to others. An example of such more readily identifiable data is a set of household survey records that, although stripped of names and addresses, contains codes for small geographic areas.

The reason for confidentiality pledges and for stringent procedures to prevent disclosure is that they improve the quality of data collected from individuals, households, and firms. It is essential that respondents believe they can provide accurate, complete information without any fear that the information will be disclosed inappropriately. Indeed, if the information was disclosed, harm might come to an individual respondent. Many government-sponsored surveys ask about sensitive topics (e.g., income or alcoholic beverage consumption), as well as about stigmatizing and even illegal behavior. The disclosure of such information might subject a respondent to loss of reputation, employment, or civil or criminal penalties. Furthermore, the breach of a confidentiality pledge would violate the principle of respect for those consenting to participate in research, even if the disclosure involved innocuous information that would not result in any social, economic, legal, or other harm (see National Research Council, 2003b:Ch.5).

The occurrence of a breach also threatens the research enterprise itself, because concerns about privacy and confidentiality are among the reasons often given by potential respondents for refusing to participate in surveys, and those concerns have been shown to affect behavior as well. Any confidentiality breach that became known would be likely to heighten such concerns and, correspondingly, reduce survey response rates. Efforts to increase researchers' access to data must, therefore, take into account the need to avoid increasing the actual and perceived risks of confidentiality breaches.

This chapter begins by reviewing research linking survey nonresponse to concerns about confidentiality. The rest of the chapter discusses some of the ways in which confidentiality breaches might occur, with special attention to how increasing access might increase both the actual and perceived risks of confidentiality breaches. Although much of this report focuses on statistical disclosure—re-identification of respondents or their attributes by matching survey data stripped of direct identifiers with information available outside the survey—these sections serve as a reminder that statistical disclosure is by no means the only, and perhaps not even the most important, way in which confidentiality breaches might occur. They also serve as a reminder that public perceptions that personal data are being misused may be as potent a deterrent to participation by potential survey respondents as an actual breach of confidentiality.

CONFIDENTIALITY CONCERNS AND NONRESPONSE IN CENSUSES AND SURVEYS

The first experimental demonstration that confidentiality concerns increase refusal to participate in a government survey comes from a National Research Council study sponsored by the U.S. Census Bureau in the late 1970s (National Research Council, 1979), but most of the evidence comes from a series of surveys commissioned by the Census Bureau in the 1990s. In the 1990 census, for example, people who were concerned about confidentiality and saw the census as an invasion of privacy were significantly less likely to return their census form by mail than those who had fewer privacy and confidentiality concerns (Singer, Mathiowetz, and Couper, 1993; Couper, Singer, and Kulka, 1998). Although such attitudes explained a relatively small proportion of the variance in census returns (1.3 percent), this proportion represented a significant number of people who had to be followed up in person to obtain information required for the census.

Analysis of the mail returns of a sample of respondents in the 2000 census yielded similar results. Once again, respondents with greater privacy and confidentiality concerns were less likely to return their census forms by mail. The variance in census returns explained by attitudes toward privacy and confidentiality was very similar to that obtained in 1990 (Singer, Van Hoewyk, and Neugebauer, 2003). In 2000, respondents with greater privacy and confidentiality concerns were also significantly less likely to provide an address to Gallup survey interviewers for the purpose of matching their survey responses to the file of census returns, and they were much less likely to respond to a question about their income.

Another way of looking at the effect of confidentiality concerns is to look at the relationship between beliefs that the census may be misused for law enforcement purposes and the propensity to mail back the census form. Of the 478 respondents in the Gallup survey following the 2000 census who believed that census data are used for none of three purposes (identifying illegal aliens, keeping track of troublemakers, and using census answers against respondents), 86 percent returned their census form by mail. The percentage dropped to 81 percent among those who selected exactly one of the three items (N = 303), to 76 percent among those who selected exactly two items (N = 255), and to 74 percent among the 171 respondents who selected all three items (Singer, Van Hoewyk, and Neugebauer, 2003). In 1990, census return rates declined from 78 percent to 55 percent on a similar index of confidentiality concerns (Singer, Mathiowetz, and Couper, 1993). Given the cost of obtaining census information that is not sent by mail, this reduction in the likelihood of returning the census form has significant consequences. Other research on the

2000 census is in accord with these findings: one study (Hillygus et al., 2006) concludes that the census return rate in 2000 would have been approximately 5 percent higher if there had not been public anxieties over privacy and what was characterized in the media and by some political leaders as unwarranted "intrusiveness."

There is also indirect evidence that requests for information on the census form that respondents consider sensitive leads to higher nonresponse rates for both the sensitive item and the entire questionnaire. For example, a 1992 experiment involving the Census Bureau's request for Social Security numbers led to a decrease of 3.4 percent in the return of the census form and an increase of 17 percentage points in the number of questionnaires returned with missing data (Dillman, Sinclair, and Clark, 1993). An experiment involving a request for Social Security numbers conducted during the 2000 census led to an almost identical result (Guarino, Hill, and Woltman, 2001:17).

Of particular interest in this context is the finding that concerns about confidentiality and negative attitudes toward data sharing increased substantially between 1995 and 2000 (Singer et al., 2001:Tables 2.16-17, 2.21-29). People's stated willingness to provide their Social Security numbers also declined, from 68 percent in 1996 to 55 percent in 1999 (Singer et al., 2001:Table 2.45). Several studies (summarized in Bates, 2005) have also documented that it has become increasingly difficult for the Census Bureau to obtain Social Security numbers. In the Survey of Income and Program Participation, there was an increase in refusals to provide them from 12 percent in the 1995 panel to 25 percent in the 2001 panel; in the Current Population Survey, there was an increase in refusals from approximately 10 percent in 1994 to almost 23 percent in 2003.

Evidence about the effects of concerns about privacy and confidentiality on response to nongovernmental surveys is provided by a series of small-scale experiments carried out in the context of the Survey of Consumer Attitudes (SCA). The SCA is a national telephone survey fielded every month at the University of Michigan, primarily to measure economic expectations and attitudes.

The first experiment, conducted in 2001, was designed to investigate what risks and benefits respondents perceived in two specific surveys—the National Survey of Family Growth (NSFG) and the Health and Retirement Study (HRS)—and how these perceptions affected their willingness to participate in the research. After hearing the description of each study, respondents were first asked whether or not they would be willing to take part in the survey, and if not, why not; they were then asked whether or not they thought each of several groups (family, businesses, employers, and law enforcement agencies) could gain access to their answers and how much they would mind if they did. Both the perceived risk of disclosure (how likely various groups were seen as gaining access to respondents' answers along with their names and addresses) and the perceived harm of disclosure (how much respondents would mind such disclosure) significantly predicted people's willingness to participate in the survey described. Perceived benefits, as well as the ratio of risk to benefit, were also highly significant.

In January and April 2003, two virtually identical experiments were carried out, again on the SCA (Singer, 2004). The introductions to both surveys mentioned the possibility of record linkage—medical records in the case of NSFG and government (financial) records in the case of HRS. Respondents who indicated that they would not be willing to take part in the survey described (48 percent of the sample) were asked why they would not do so. The most frequent reasons given—59 percent of all first-mentioned reasons—were that the surveys were too personal or intrusive or that they objected to giving out financial or medical information or providing access to medical or financial records. As in the previous experiment, perceptions of disclosure risk, disclosure harm, individual and social benefit, and the ratio of risk to benefit were strong and significant predictors of people's willingness to participate. Similarly, an experiment in connection with the 2000 census found that respondents primed to consider privacy issues had higher rates of item nonresponse to census long-form questions than a control group (Hillygus et al., 2006).

These experiments point to the importance of *perceptions* of disclosure risk, as well as of actual risks. Public awareness of confidentiality breaches in nongovernment surveys may adversely affect perceptions of the risks arising from participation in government surveys. That is, public knowledge of a breach of confidentiality by an employee of a government benefit agency or private insurance company may increase concern about such breaches by federal statistical agencies, such as the Census Bureau. Similarly, public knowledge of legal demands for identified records, such as subpoenas for data about individuals by law enforcement agencies or attorneys for plaintiffs or defendants, may increase such concerns. Similar concerns and effects may result from identity theft, through unauthorized access to an individual's credit card account and Social Security numbers; from misuse of medical records by entities (e.g., insurance companies) that are entitled access to them for administrative purposes; or from misuse of administrative records or survey records by employees of a data collection agency. And, as noted above, such concerns about confidentiality adversely affect the likelihood of participation in government surveys.

WHY CONFIDENTIALITY BREACHES MIGHT OCCUR

Carelessness and Illegal Intrusions

Survey researchers have identified various ways in which the confidentiality of individual respondents might be breached. Perhaps the most obvious and common threat to confidentiality protection of research data arises from simple carelessness—not removing identifiers from questionnaires or electronic data files, leaving cabinets unlocked, not encrypting files containing identifiers, talking about specific respondents with others not authorized to have this information. Although there is no evidence of respondents having been harmed as a result of such negligence, it is important for government data collection agencies and private survey organizations to be alert to these issues, provide employee guidelines for appropriate data management, and ensure that the guidelines are observed.

Confidentiality may also be breached as a result of illegal intrusions into the data. For example, in 1996, ten Social Security employees (bribed by outsiders) were found to have stolen confidential information from agency computers. The key piece of information was mothers' maiden names, which were stored in a database with password protection but less stringent security than that protecting earnings statements and other private information. The information was used to activate credit cards of residents in the New York area. Identity theft has been increasingly in the news since then.

As detailed data collected under a pledge of confidentiality are increasingly made available to researchers through licensing agreements or in research data centers, the potential for inadvertent disclosure as a result of carelessness and through deliberate illegal intrusions may also increase unless strong educational and oversight efforts accompany such means of access. In Chapter 5 we offer several recommendations designed to strengthen protections against these sources of disclosure of information about individuals.

However, the extent of the problem is not easily determinable, either by assessing past experience or predicting future effects. Numerous media stories have documented harms of identity theft from such sources as credit card and banking data. In contrast, there is no documented evidence of harms from misuse of research data or carelessness by researchers or others. Overall, very little is known about how many breaches of confidentiality may actually occur in such settings or how many people are harmed as a result. Under most circumstances, attempted breaches are difficult to detect, and relying on self-reports is problematic. A July 1993 survey by Harris, for example, reported that between 3 percent and 15 percent of the public, depending on the person or organization asked about, believed that medical information about them had ever been improperly disclosed, and about one-third of these said they had been harmed by the disclosure (Singer, Shapiro, and Jacobs, 1997). But the accuracy of these reports is unknown. Moreover, disclosure of medical information to an insurance company may be permitted by law but regarded by survey respondents as improper. For many people, questions about breaches of confidentiality may be highly abstract so that their ideas about the uses that might be made of their medical information are limited. As a result, little is really known about what people have in mind when they answer such questions, and even less about the actual state of affairs. Again, in Chapter 5 we offer some recommendations to address this concern.

Law Enforcement and National Security

Potentially more serious threats to confidentiality than simple carelessness are legal demands for identified data, which may come in the form of a subpoena or as a result of a Freedom of Information Act (FOIA) request. Requests may also come from a law enforcement or national security agency to a statistical or other government agency; the legal status of such requests is not fully resolved, as discussed below. Individual records from surveys that collect data about such illegal behaviors as drug use are potentially subject to subpoena by law enforcement agencies. To protect against this possibility, researchers and programs studying mental health, alcohol and drug use, and other sensitive topics, whether federally funded or not, may apply for certificates of confidentiality from the U.S. Department of Health and Human Services. The National Institute of Justice (in the U.S. Department of Justice) also makes confidentiality certificates available for criminal justice research supported by agencies of the U.S. Department of Justice. Such certificates, which remain in effect for the duration of a study, protect researchers in most circumstances from being compelled to disclose names or other identifying characteristics of survey respondents in federal, state, or local proceedings (42 *Code of Federal Regulations* Section 2a.7, "Effect of Confidentiality Certificate"). The confidentiality protection afforded by certificates is prospective; researchers may not obtain protection for study results after data collection has been completed.

Protection for identifiable statistical data collected by federal agencies or their agents under a promise of confidentiality is also provided by the Confidential Information Protection and Statistical Efficiency Act (CIPSEA), which was enacted as Title V of the E-Government Act of 2002 (P.L. 107-347). The legislation is intended to "safeguard the confidentiality of individually identifiable information acquired under a pledge of confidentiality for statistical purposes by controlling access to, and uses made of, such information." The statute includes a number of safeguards to ensure that information acquired for statistical purposes under a pledge of confidentiality "shall be used by officers, employees, or agents of the agency exclusively for statistical purposes," and "shall not be disclosed by an agency in identifiable form, for any use other than an exclusively statistical purpose, except with the informed consent of the respondent." Identifiable information can be disclosed, under proper conditions, for "statistical activities," which are broadly defined to include "the collection, compilation, processing, or analysis of data for the purpose of describing or making estimates concerning the whole, or relevant groups or components within, the economy, society, or the natural environment" as well as "the development of methods or resources that support those activities, such as measurement methods, models, statistical classifications, or sampling frames."

CIPSEA also imposed additional responsibilities on statistical agencies, requiring them to "clearly distinguish data or information [they collect] for nonstatistical purposes," and to "provide notice to the public, before the information is collected, that the data could be used for nonstatistical purposes." Nonstatistical purposes are defined as "any administrative, regulatory, law enforcement, adjudicatory, or other purpose that affects the rights, privileges, or benefits of a particular identifiable respondent" and include disclosure under the Freedom of Information Act. The act also provides criminal penalties for a knowing and willful breach of confidentiality by employees of the sponsoring agency and any of its "agents," who may be data collectors or outside analysts.

CIPSEA offers great promise for increasing researcher access to confidential data. Fulfillment of that promise requires, in the first place, coordination of access and protection procedures across the various agencies in order to satisfy the uniform protection promised by the act. At the time of this report, the Office of Management and Budget is preparing regulations to implement the safeguards under CIPSEA. These implementing regulations will be critically important in translating a statutory right into clear rules that protect research participants across all federal agencies. The regulations are expected to define both the reach of protection for confidential statistical records and the opportunity for research access.

The regulations will have to cover a wide range of questions, such as:

- Other than federal agency personnel, who can qualify as an "agent" under the statute and thereby be eligible for research access to identifiable records?

- Does a licensing agreement between an agency and a private researcher for research access fall within the coverage of the statute?
- What degree of risk of inadvertent disclosure of identifiable information will govern the release of anonymized records?
- What form of public notice is required when a statistical agency collects identifiable information for nonstatistical purposes?
- Which, if any, of the CIPSEA protections extend to identifiable administrative records that are used for research purposes?
- How does CIPSEA affect existing regulations and practices under other agency statutes that protect research records?
- What procedural safeguards are required to monitor the work of agency staffs and nonagency personnel who are deemed "agents" under CIPSEA?

Fulfillment of the potential for research access to data sharing under CIPSEA will ultimately also require companion legislation that would permit the Census Bureau to share tax information that it receives from the Internal Revenue Service (IRS) with the Bureau of Labor Statistics and the Bureau of Economic Analysis in order to reconcile the business lists built by the three agencies. In the absence of such legislation, data sharing for research among the three agencies is restricted to information that does not include, or derive from, tax data. Another topic that may need future legislative attention is the sharing of individual data, since the data-sharing provisions of CIPSEA currently apply only to business data.

The seeming clarity of the protections afforded by CIPSEA is clouded by concerns about potential conflict with access to identifiable data for national security purposes. In the past, government agencies have attempted to use confidential data collected by a statistical agency for law enforcement purposes, especially in times of heightened national security concerns. Seltzer and Anderson (2003) review attempts by various government agencies to obtain confidential census data between 1902, when the Census Bureau was established as a permanent agency, and 1965. A few of these attempts in the years before enactment of Title 13 in 1929—especially those involving national security—were successful and, in at least some of them, actual disclosure of information about individuals for national security or law enforcement purposes occurred. In 1917, for example, personal information from the 1910 census was released to courts, draft boards, and the Justice Department for several hundred young men suspected of not complying with the draft (Barabba, 1975:27, cited in Seltzer and Anderson, 2003). During World War II, according to Prewitt (2000:1): "The historical record is clear that senior Census Bureau staff proactively cooperated with the internment [of Japanese Americans], and that census tabulations were directly implicated in the denial of civil rights to citizens of the United States who happened also to be of Japanese ancestry."[1] In 2004 the Census Bureau provided information about the residences of Arab Americans to the Customs and Border Protection agency of the U.S. Department of Homeland Security, but that information was also available on a public-use site and involved data masked to protect confidentiality. Although this incident was not a violation of law, it was perceived as such by many people, as well as a violation of trust (see Clemetson, 2004).

The 2001 USA Patriot Act, which is being considered for renewal by Congress as this report is being written, includes provisions for access by the U.S. Attorney General to identifiable research records of the National Center for Education Statistics (in the U.S. Department of Education). This provision appears to be unique: the panel is not aware of any other provisions for access to confidential research data for national security purposes. Both the Homeland Security Act of 2002 (P.L. 107-296) and the Intelligence Authorization Act for Fiscal Year 2003 (P.L. 107-306) make clear that exchange of federal agency information for homeland security needs does not include exchange of individually identifiable information collected solely for statistical purposes. Nevertheless, as Seltzer and Anderson have shown, national security crises have in the past led to circumventions or actual violations of confidentiality guarantees.[2]

Statistical Disclosure

Breaches of confidentiality due to carelessness, as well as those from illegal intrusions, are obviously more likely to occur if a data file contains direct identifiers—name, address, or Social Security number, for example. Yet there is increasing awareness that even without such identifiers, statistical disclosure may be possible. "Statistical disclosure" refers to the re-identification of respondents to a survey (or their attributes) even though direct identifiers such as names and addresses have been removed from the data file. Statistical disclosure involves using data available outside the survey to breach the protection thought to have been

---

[1] In that same speech, former Census Bureau Director Kenneth Prewitt apologized on behalf of the agency for its activities in connection with the internment of Japanese Americans. For a detailed history of Census Bureau cooperation with national security activities during World War II, see Seltzer and Anderson (2000).

[2] Although it is not directly relevant to national security, the Shelby Amendment (part of P.L. 105-277) and the Data Quality Act (see Chapter 2) also have implications for confidentiality protection that have not yet been fully determined.

afforded a survey data set by various data deletion and masking techniques. Re-identification of respondents may be increasingly possible because of high-speed computers, external data files containing names and addresses or other direct identifiers as well as information about a variety of individual characteristics, and sophisticated software for matching survey and other files. In Chapter 2 we noted some of the factors that may increase statistical disclosure risk and harm for respondents in government-sponsored surveys, including factors that are integral to the survey design and factors that are external to data collection agencies and researchers. In addition, there is a growing concern by data collection agencies (see below) that wider dissemination of research data may itself increase disclosure risk.

For a breach of confidentiality due to statistical disclosure to occur, there must be the technical or legal means, as well as the motivation to use them. With regard to motive, there are (at least) four: curiosity, sport (e.g., hackers), profit (e.g., identity theft), and law enforcement or national security.[3]

Breaches occurring because of curiosity or sport may never become known to the respondent. However, the confidentiality pledge has been violated, and ethical harm has been done, even if all that has happened is that someone has identified a record in a data file and not used it for any purpose.

The further harm a breach of confidentiality may cause depends in part on the type of intruder and the type of data. Federal regulations for the protection of human subjects of research (in the Common Rule, 45 *Code of Federal Regulations* 46) focus mainly on the potential harm to an individual's reputation, livelihood, or liberty resulting from the disclosure of confidential information, suggesting that disclosure of deviant or illegal behavior or unpopular beliefs is most likely to be harmful. However, if an intruder's aim is identity (or property) theft, then anything that permits the appropriation and abuse of another's identity may be harmful to that individual. If the intruder is a hacker simply out to embarrass the survey organization, then public identification of one or more survey participants may be enough to do harm to the data collection and research enterprise, even if the information is not sensitive and the participants are not directly harmed.

A survey design factor that, prima facie, would seem to increase the risk of statistical disclosure is the increasing number and diversity of at-

---

[3]Ochas et al. (2001) list additional reasons why reidentification might be attempted: investigative reporting, blackmail, marketing, denial of insurance, and political action.

tributes asked about and stored on the data record for each respondent. The greater the number of attributes about which information is provided, the greater is the theoretical potential for re-identification. Since the late 1960s, surveys have become more detailed on several dimensions. Thus, more and more surveys are collecting detailed socioeconomic attributes for individuals and households; more and more surveys are asking about individual behaviors, including those that are risky and even illegal; and more and more surveys are longitudinal in design, collecting repeated measurements on the same individuals. In addition, a growing number of both cross-sectional and longitudinal surveys collect data about an individual from multiple sources: for example, surveys of children in which data are obtained from parents, schoolteachers, and others, and surveys that collect information about individuals, the schools they attend, and the neighborhoods in which they live.

More recently, a small but growing number of surveys are making use of new technologies for collecting biological and geographic information, which in turn make it easier to identify respondents—or more difficult to conceal their identity (see, e.g., National Research Council, 1998, 2001a). Such information, which includes DNA samples, biological measurements, and geospatial coordinates, complicates the problem of making data files anonymous and heightens the dilemma of data collection agencies and researchers who want to increase access to the data they collect while protecting the confidentiality of respondents (see, e.g., Abowd and Lane, 2004).

Other factors that may increase the risk of statistical disclosure are external to the survey organization and researcher. As noted above and in Chapter 2, these factors include the increasing availability of files in the external environment that are suitable for matching to survey records and, in addition, contain names and addresses or other direct identifiers; the ready availability of matching software; and quantum increases in the processing and storage capabilities of computer hardware and software, which make it possible to manipulate multiple files with rapidity and relative ease. If microdata have been stripped of direct identifiers but no added steps have been taken to minimize disclosure risk, it is relatively easy to match the file with external databases that contain some of the same variables as the original midcrodata (plus names and addresses) and thus to identify some respondents (see, e.g., Winkler, 1988). Similar research has been conducted by others (see, e.g., Sweeney, 2001). However, the panel knows of no information on whether this has been done other than in a research context.

Statistical agencies and survey organizations understandably worry that wider access to ever more complex datasets, in an era of cheap, capacious computing technology and many outside data sources for matching, will increase the risk of statistical disclosure and the potential for harm to respondents, as well as to survey participation. Although many factors seem to increase the risk of disclosure, there is some evidence suggesting that increasing the number of attributes in a data record does not necessarily lead to increased disclosure. For example, the Retirement History Survey (RHS), which followed people who were aged 58-63 in 1969 for 10 years, made more information publicly available than the HRS, which has followed people aged 51 and older since 1992. Yet there are no known instances of a breach of confidentiality for the RHS, from which microdata have been publicly available for more than 30 years. Similarly, there are no known instances of disclosure or consequent harm for other richly detailed and long-available datasets, such as the Panel Study of Income Dynamics, which has followed families and their descendants for more than 35 years. Although this evidence is suggestive, it is important for statistical and other agencies to know how often inappropriate disclosures of information actually occur and what the risk of disclosure is in different circumstances.

Ultimately, decisions about how much disclosure risk is acceptable in order to achieve the benefits of greater access to research data involve weighing the potential harm posed by disclosure against the benefits potentially foregone, as well as a judgment about who should make those decisions. The panel does not resolve these difficult issues. Rather, in Chapter 5 we recommend research to reduce disclosure risk while preserving data utility. We also recommend research that improves estimation of disclosure risk and procedures for monitoring the actual frequency of disclosure. Finally, we recommend continuing consultation with data users and data providers about all of these issues.

5

Reconciling the Benefits and Risks of Expanded Data Access

The charge to this panel—and the challenge to those who collect data from individuals and organizations and those who use them—is to understand and weigh the tradeoffs between the benefits and risks of increased access to research data. The benefits of increased access are better data for policy analysis and research; the risks are breaches of confidentiality and their consequences. As noted in Chapter 4, breaches of confidentiality can occur in a variety of ways. The work of this panel has focused primarily on statistical disclosure—the re-identification of individual respondents (or their attributes) through the matching of survey data with information available outside the survey.

Achieving the benefits of access to research data presupposes a willingness by people in households and organizations to provide detailed and sometimes sensitive information for government-sponsored statistical surveys and censuses. Such willingness requires public trust: trust that the data will be used for important research and policy purposes and that the confidentiality of the information will be maintained. Thus, the agencies that collect data have an obligation to communicate clearly to respondents the purposes of the data they collect and to assure respondents of the confidentiality of the information they provide.

In this final chapter of the report, we draw on technological, legal, administrative, and statistical sources to offer recommendations that we believe will facilitate access to research data while protecting the confidentiality of information provided by the public. We offer recommendations in eight broad areas: documenting the use of research data; planning for access to data through a variety of modes; expanding access to public-use files; facilitating access to research data centers; expanding remote access capabilities; broadening the use of licensing and bonding agreements; assuring informed consent; and safeguarding confidentiality through training, monitoring, and education in research ethics.

DOCUMENTING USE

As discussed in Chapter 3, data collected by government agencies benefit society by providing the basis for research and policy analysis, which, in turn, can inform policy makers and the public. Longitudinal surveys that obtain data for analyzing the determinants and consequences of social and economic behaviors have been a major positive development for research and policy in the past 30 years. Linking survey and administrative data can create particularly rich datasets that, in some cases, can substitute for additional surveys, thus reducing respondent burden as well as government costs. To realize the full potential of these data, researchers outside of government need access to them.

The United States has a decentralized, pluralistic research structure, in which not only the staffs of statistical and other data collection agencies, but also researchers in many institutions, with different policy preferences and perspectives, can and should have access to data. This broad access by independent researchers and analysts provides checks and balances on the government's dominant role in public policy. The fruits of the diverse, decentralized social science research enterprise include not only studies that contribute to longer-term understanding of the dynamics of individual, household, and organizational (e.g., business) behavior, but also analyses that contribute directly to public discourse and the development and evaluation of public policies.

The rapid expansion of information and communication technologies in the past decade has enhanced the potential for society to benefit even more from the collection and provision of data for social science research. With greater computing power and better software, the value of data, especially complex longitudinal microdata, has increased. With such data, investigators can estimate and test models that are closer approximations to reality and that directly address problems of causal inference. Although the new technologies make it more difficult to protect the confidentiality of respondents, they also enhance the possibilities for disseminating research data more widely. Failure to use such technologies inhibits the ability of society to exploit fully the rich data collected by federal agencies or others on their behalf.

In part because of the public goods aspect of data collection and research and in part because of the decentralized structure of both data collection and data analysis activities, it is difficult for data collection agencies, research organizations, or society to assess the value of the data produced. Although the benefits of data access are compelling (see Chapter 3), no one has developed a generally accepted approach for quantifying the extent and value of that access or placing a quantifiable value on the uses of the data.

Careful tracking of the numbers and types of users and the body of research produced would provide a sense of the importance of various data products. More broadly, we conclude that a more comprehensive record of the research use of data would be valuable to agencies, policy makers, and researchers. Registries and documentation of data use could help foster understanding by both the public and policy makers of the value of various kinds of data, including microdata, and the research that these data inform.

In addition to a record of the use of research data, it is important to know how many requests for data are received and how many of those are denied, as well as the time required to gain access when the request is granted. The panel that produced *Private Lives and Public Policies* explicitly recommended that procedures be established for keeping records of data requests denied or partially fulfilled (National Research Council, 1993:100). This panel endorses that recommendation, which has not yet been implemented. Records of such requests for confidential data may also be useful to agencies in monitoring confidentiality protection procedures and actual breaches of confidentiality (see "Research on Breaches of Confidentiality," in this chapter).

A first step in documenting use would be to assemble bibliographies of research papers for particular data sets. The bibliographies of research papers maintained for some existing datasets offer models of the kinds of documentation that we envision. They include: the National Longitudinal Surveys of Labor Market experience (NLS, housed at the Center for Human Resource Research at Ohio State University), the Panel Study of Income Dynamics (PSID, housed at the University of Michigan), the Health and Retirement Study (HRS, housed at the University of Michigan), and the General Social Survey (GSS, housed at the University of Chicago). The U.S. Census Bureau has a bibliography of nearly 2,000 references to published and unpublished work using data from the Survey of Income and Program Participation (SIPP) and the predecessor Income Survey Development Program, but the bibliography has not been updated since 1998 (www.sipp.census.gov/sipp/aboutbib.html). In addition to research publications, such as articles and books, we encourage bibliographies of research presentations at scholarly meetings and any research analysis presented in the form of software applications.

Federal statistical agencies could use such bibliographies, not only to assess the use of their data sets, but also as a sampling frame for contacting researchers to obtain feedback on the quality and usefulness of the data. Similarly, research funding agencies could usefully commission analyses that build on research bibliographies to assess the extent, quality, and importance of social science research conducted with statistical data, particularly microdata.

> **Recommendation 1** As a first step to facilitate systematic study of the extent and value of data access for research, public and private agencies that collect social science research data should maintain up-to-date bibliographies of research and policy analysis publications, presentations, and software applications that use the data.

ACCESS THROUGH MULTIPLE MODES

The actual data collected for statistical purposes from households, individuals, business establishments, and other organizations through censuses and surveys under a pledge of confidentiality are never made available to users. Instead, data are made available either in the form of confidential, restricted-access data files or in the form of anonymized data products, including published tables and microdata files.

Confidential files delete direct identifiers such as names and addresses but retain the observational structure of the original data and include all of the value added by an agency to generate its published statistics (such as analysis weights, imputation for unit and item nonresponse, data quality edits, geocoding, industry coding, occupation coding). They also contain details (such as place of residence, occupation, industry, income, and wealth) that cannot be made available on public-use files. In some cases an agency may also create links to administrative records or other data files.

Public-use microdata files, constructed from the confidential files, contain data that have been masked through various steps (rounded, aggregated, edited) or that have been altered through such techniques as multiple imputation to ensure that individual respondents and their attributes cannot be identified. Public-use files are the most accessible and widely used microdata products made available by statistical agencies, but their value for much policy-relevant research is limited. To exploit the full research and policy value of microdata, researchers will often need access to the confidential files. Modes of access to such data (restricted access modes) include access at supervised locations, remote access with prior review of data output, and access through licensing and bonding agreements (discussed below).

Because public-use files are available to all, statistical agencies must exercise great care to ensure their anonymity, as well as their usefulness and accuracy. The ultimate goal is to provide public-use data that will yield the same statistical inferences that would be derived from the confidential data. But since many of the analyses that will be performed on public-use files cannot be foreseen at the time of their release, the process of assuring their quality requires a continual feedback relationship between the public-use files and the underlying confidential data. Analyses performed on the confidential data can be used to improve the next generation of public-use data in several ways. For example, such analyses can identify errors in the original data, the public-use data, or both, as well as anomalies that should be corrected in statistical procedures (such as imputation). They can also identify ways in which the public-use data could be made more useful for research by altering the procedures for confidentiality protection for some items. Such analyses can also suggest priority areas for improvement and enhancement of data content in subsequent data collections.

This feedback relationship serves two related purposes. First, it improves the quality and relevance of public-use files, an outcome of great importance to the statistical agencies because they must attest to the usefulness of these products as a source of statistical information to a broad audience. Second, it justifies a substantial investment in facilitating access to the underlying confidential microdata because such an investment supports an agency's core mission of assuring the quality and usefulness of its public-use products. Both kinds of access—restricted access to the confidential data and unrestricted access to inference-valid public-use data— are needed, not only to accommodate different types of users with different purposes, but also to maintain and improve the quality of public-use files and to obtain accurate estimates of the error entailed by their use.

Economists and statisticians have begun to model the optimum mix of different data access modes by examining the costs and benefits of providing research access through public-use data products or through restricted access modes (Abowd and Lane, 2004). Such modeling is in its infancy but could be valuable at several levels. For individual variables, cost-benefit modeling might identify specific items that could be moved from restricted access to public-use data without impairing confidentiality protection and, conversely, items that should be moved from public-use products to restricted access modes.

At a broader level, cost-benefit modeling could be used to evaluate the tradeoffs among the various forms of restricted access—research data centers, remote access, licensing—as well as among different ways of restricting data (through various masking techniques and various ways of producing synthetic data). Cost-benefit modeling entails the use of a large number of assumptions, including the expected number, variety, and value of uses in each mode; the disclosure risk and associated costs for each mode; the user costs for access through a research data center compared with a public-use file or a licensing arrangement; and the producer costs for preparing the public-use product compared with running a research data center and approving and overseeing licensees. For realistic estimation, cost-benefit modeling will require empirical estimates of such factors as disclosure risks and costs, numbers and benefits of research uses, user costs of access for various modes, and producer costs of providing various modes of access. Such estimates do not currently exist, and some of them, including disclosure risks and costs and the benefits of research use, are not easy to develop, although promising work is under way on estimating disclosure risk (see Reiter, 2003; see also "Public-Use Data" in this chapter).

Given the importance of facilitating research access to statistical microdata, statistical and research agencies should encourage research on the most efficient allocation of resources among different access modes to guide their planning and to support changes (including legislative changes, if needed) to facilitate one or other type of access. If a variety of agencies, as well as users, are involved in such research and planning efforts to develop data access programs, the data that agencies produce are more likely to be widely used, ultimately leading to better research and policy analysis and to important feedback to data producers that can enable them to enhance data quality and relevance.

Examples of the kinds of involvement of users and producers that we envision include the major longitudinal surveys that are funded through grants to academic survey organizations (including the HRS and the PSID), which have active boards of users and potential users that guide their development. Although most federal statistical agencies also have outside advisory groups, they rarely focus on data access programs for particular data sets. However, one example of focused user involvement for a statistical agency microdata collection is the Association of Public Data Users Working Group on SIPP Data Products, which was active from 1989 to 1994. It played a major role in the development of more user-friendly data products and comprehensive user documentation from the Survey of Income and Program Participation.

> **Recommendation 2** Data produced or funded by government agencies should continue to be made available for research through a variety of modes, including various modes of restricted access to confidential data and unrestricted access to public-use data altered in a variety of ways to maintain confidentiality.

Recommendation 3 The National Science Foundation, the National Institutes of Health, and major statistical agencies should support research to guide more efficient allocation of resources among different data access modes.

Recommendation 4 Statistical and other data collection agencies should involve users more fully in planning modes of access to their data.

PUBLIC-USE DATA

Improving Quality

Public-use files, introduced in the 1960s, are the most widely available form of research data. As described above, information generated from investigating confidential data in a restricted access environment can be used to improve their quality and relevance. Improved public-use data files, in turn, can reduce the need and demand for restricted access for some data sets. If, for example, it is possible to move certain variables, such as summary measures or average values, from restricted-access to public-use files while maintaining confidentiality protections, overall access to research data could be increased at reduced cost to users and producers. Doing so requires further sustained research and development of methods that permit data collection agencies to assess the increase in disclosure risk posed by the addition of specific variables to existing public-use files.

Data from the HRS provides an example of the kind of assessment needed. The primary insurance amount (PIA), which is calculated from the detailed Social Security earnings records of HRS respondents, is a very desirable variable for many researchers, most of whom do not need the detailed Social Security administrative records. PIA information would seem to present little additional disclosure risk because many patterns of lifetime earnings will produce the same PIA. Thus, one cannot recover any particular detailed earnings history from a single PIA number. But because HRS policy requires that any variable derived from Social Security records must be a confidential variable, all researchers must apply for access to the confidential files (a tedious and costly process) in order to gain access to the PIA data. Research that demonstrated the absence of increased disclosure risk might make it possible to alter the HRS policy so that PIA data could be made available as part of the public-use file. Such a step would be likely to substantially reduce the demand for detailed earnings data. Moreover, and paradoxically, by increasing access to the public-use file, the overall risk of disclosure might be reduced because fewer people would have access to the confidential data.

Other improvements to public-use files also require research. Currently, these files rely on a variety of disclosure limitation methods, such as rounding, coarsening, data swapping, and top coding (see Duncan, 2002). Research is needed to examine and quantify the tradeoffs between disclosure risk and data utility when these methods are used. Since all public-use files require disclosure limitation in order to protect confidentiality, it is crucial that the methods used maximize both the utility of the data and the protection of confidentiality, recognizing that a zero risk of disclosure can never be guaranteed.[1]

Disclosure risk research has begun to advance from the delineation of general approaches with simulated data to empirical estimation of the disclosure risk in existing microdata sets. For example, Reiter (2003) estimated changes in disclosure risk by altering variables, such as age and property taxes, in the Current Population Survey, under alternative assumptions about what a data snooper might know (see also Duncan et al., 2001; Duncan and Stokes, 2004). Much could be learned from more empirically based work with different variables and different files, including work by agencies that do not currently release much, if any, public-use microdata, such as the Social Security Administration. That work might determine that such disclosure limitation methods as calculating summary measures or average values from confidential data and attaching these measures to other commonly used microdata could result in highly useful, fully protected public-use microdata for research and policy analysis on such important topics as taxation and retirement income security.

Currently, studies of the probabilities of disclosure include estimates of the technical possibility of matching public-use survey data with other widely available information, but they do not include estimates of the likelihood that such matching would be attempted. Consequently, estimates of disclosure risk may overstate the potential risk. Yet without data on people's propensities for snooping through different types of survey data (which may not be the same as their propensities for hacking into credit card company records or other data sets that present clear financial incentives), one cannot empirically estimate the likelihood of attempts to re-identify records in particular microdata sets. However, it would be possible to use different assumptions about those propensities to help agencies set bounds on their estimates of disclosure risk for particular types of data.

---

[1]For a bibliography on disclosure risk analysis and best practices for statistical disclosure control, see O'Rourke and Gutmann (2005).

Another factor that could be considered in the calculus of the disclosure risk-benefit tradeoff is the degree of harm that might result from disclosure of particular types of information about an individual or organization. It may be that some degree of uncertainty about the risk of disclosure could be tolerated for data that are not likely to put an individual at risk of serious harm, while a worst-case assumption about disclosure risk would be prudent for highly sensitive data (e.g., reports of illegal drug use or detailed information about financial assets). In making a decision about acceptable levels of disclosure risk, it is important that data providers' views be considered along with those of the data collection agency and potential research users (see below).

Research on disclosure limitation methods should include establishment (business) data as well as household and individual data. This topic has so far received relatively little attention for business data because of the presumption that establishments are too easy to identify unless their attributes are so heavily masked that the resulting public-use microdata would have little analytic value. Yet there may be methods that could be effective in protecting confidentiality and preserving research utility for establishment-based public-use microdata. Without research on different methods, access to establishment data is likely to continue to be severely restricted, limiting research and policy analysis on important topics.

Methods of disclosure limitation based on synthetic or virtual data, which are constructed from confidential data through partial or complete multiple imputation techniques, show promise in safeguarding confidentiality and permitting the estimation of complex models; they should continue to be explored as an alternative to other disclosure limitation methods (see, e.g., Abowd and Woodcock, 2001; Doyle et al., 2001; Raghunathan; 2003). The chief drawback of synthetic data, as of masked data, is their potential for yielding misleading results, especially when complex models are estimated. Empirical estimates of the amount of error introduced by imputation-based disclosure limitation methods under various assumptions are needed both to demonstrate their utility and to suggest ways in which they can be improved. Such research, which should also be conducted for perturbation and data swapping methods, will become increasingly important as highly useful but highly sensitive variables—such as biological markers, including DNA samples, and geospatial coordinates—are increasingly linked with survey responses.

Synthetic public-use microdata could also facilitate data access by providing a means for researchers to explore, test, and refine estimation models at relatively low cost before incurring the higher costs of access to confidential data through a research data center or another restricted access mode. For this purpose, the synthetic data would need to meet a high standard for supporting valid inference but not as high a standard as would be necessary for research and policy analysis that relied on the synthetic data alone.

Recommendation 5 Agencies that sponsor data collection should conduct or sponsor research on techniques for providing useful, innovative public-use data that minimize the risk of disclosure. Such research should also be a funding priority for the National Institutes of Health and the National Science Foundation. In particular, research should be directed to:

(1) developing measures for quantifying disclosure risk;
(2) estimating the effect on disclosure risk of adding selected variables from confidential data files to public-use files;
(3) estimating and improving the utility-disclosure limitation trade-offs of alternative disclosure limitation methods, including synthetic data; and
(4) developing disclosure limitation methods for establishment data.

Facilitating Access to Public-Use Files

Academic researchers in the United States need approval from an Institutional Review Board (IRB) to conduct research involving human participants, including, in many cases, secondary analyses. Currently many, perhaps most, IRBs lack the expertise required to review the adequacy of the confidentiality protection for research that involves original data collection. As a result, researchers spend much time justifying to IRBs proposed reanalyses of public-use microdata from federal agencies and established data archives that incorporate best practices for confidentiality protection (see National Research Council, 2003b).

The Panel on Institutional Review Boards, Surveys, and Social Science Research recommended a new confidentiality protection system, built on existing and new data archives and statistical agencies, to facilitate secondary analysis of public-use microdata (National Research Council, 2003b:138 [Recommendations 5.2 and 5.3]). Such a system would permit IRBs to exempt secondary analysis with such data from review as a matter of standard practice under clause 46.101(b)(2) of 45 CFR 46, Subpart A, Federal Policy for the Protection of Human Subjects.

The system could be developed as follows: the Office for Human Research Protections in the U.S. Department of Health and Human Services would work with statistical agencies, appropriate interagency groups, and data archives to develop a certificate to accompany the release of public-use data sets. Data producers and archives could obtain certification for all of their public-use data sets or for individual files if they rarely produce public-use data. Such a certificate would attest that the public-use file reflects good practice for confidentiality protection and that the data were collected with appropriate concern for informed consent and other human research participant protection issues. With such a certificate, the IRB would exempt from further review any analysis that proposes to use only the data from the certified files.

In supporting its recommendation, the panel noted (National Research Council, 2003b:138-139):

> We argue that IRB review of secondary analysis with public-use microdata is unnecessary and a misuse of scarce time and resources . . . If the data in a file have been processed to minimize the risk of re-identifying a respondent by using widely recognized good practices for confidentiality protection, then the research is eligible for exemption under the Common Rule. . . .

Development of the process of certification assumes heavy participation by the statistical agencies, the Office of Human Research Protections, the Office of Management and Budget Statistical and Science Policy Office, and interested data archives and would clearly require an initial investment of significant amounts of time and money. But if IRBs throughout the country accepted certification as sufficient to exempt from review research involving data from a statistical agency or a nationally recognized survey organization, access to research microdata would be considerably enhanced. We therefore endorse the recommendations of the earlier panel.

Recommendation 6 To enhance access to public-use files for secondary analysis, we endorse the recommendations of the Panel on Institutional Review Boards, Surveys, and Social Science concerning establishment of a new system of confidentiality protection for public-use microdata based on existing and new data archives and statistical agencies. Statistical agencies and participating archives would certify that public-use data sets obtained from them were sufficiently protected against statistical disclosure to be acceptable for secondary analysis, and IRBs would exempt such analyses from review on the basis of the certification provided.

Extending Legal Obligations to Data Users

At present, the obligation to protect individual respondents falls primarily on those who collect the data, thereby creating a disincentive for providing access to other researchers. We believe this obligation should be extended to the users of public-use data as well. All releases of statistical data by federal agencies, including public-use data files, should include a warning that the data are provided for research purposes only and that any attempt to identify an individual respondent in the data file is a violation of federal law and will result in penalties comparable to those currently imposed only on agency personnel and licensed users. Such a warning currently accompanies public-use records released by the National Center for Education Statistics (NCES). Although such restrictions may be difficult to enforce, especially for public-use data, the legal sanction will stand as an expression of professional norms regarding the use of research data.

The ability to seek penalties may require new legislation for most agencies. The language that is in place for data sets from the NCES (P.L. 107-279, Education Sciences Reform Act of 2002, Section 183(d)(6)) is an example of the kind of penalties that would be appropriate to invoke against users of public-use data from federal statistical agencies who breach confidentiality:

> Any person who uses any data provided by the Director, in conjunction with any other information or technique, to identify any individual student, teacher, administrator, or other individual and who knowingly discloses, publishes, or uses such data for a purpose other than a statistical purpose, or who otherwise violates subparagraph (A) or (B) of subsection (c)(2), shall be found guilty of a class E felony and imprisoned for not more than five years, or fined as specified in section 3571 of title 18, United States Code, or both.

Recommendation 7 All releases of public-use data should include a warning that the data are provided for statistical purposes only and that any attempt to identify an individual respondent is a violation of the ethical understandings under which the data are provided. Users should be required to attest to having read this warning and instructed to include it with any data they redistribute.

Recommendation 8 Access to public-use data should be restricted to those who agree to abide by the confidentiality protections governing such data, and meaningful penalties should be enforced for willful misuse of public-use data.

At noted above, new legislation would be required for some agencies to have the authority for such penalties.

FACILITATING ACCESS TO RESEARCH DATA CENTERS

One key way to provide researcher access to confidential data is through research data centers (RDCs), including the eight centers maintained by the Census Bureau and those maintained at the headquarters of the Agency for Healthcare Research and Quality and the National Center for Health Statistics, in the Department of Health and Human Services, the Bureau of Labor Statistics in the Department of Labor, and some other agencies.[2]

As noted in Chapter 2, the Census Bureau's RDCs represent an important step toward facilitating research access to confidential data; however, they are believed to be underused, and their use appears to be declining. Two of the three reasons for this trend are the length of the review process and the costs involved in doing research away from one's home institution. The third reason is a very stringent interpretation of the five criteria for approving a research project, which must demonstrate:

(1) a likely benefit to the Census Bureau under Title 13 and, indeed, that its *predominant purpose* is to provide one or more Title 13 benefits, such as improving imputations for nonresponse (see www.ces.census.gov/ces.php/guidelines);

(2) scientific merit in terms of a project's likelihood to contribute to existing knowledge (which is similar to the criterion for research-funding agencies, such as the National Science Foundation [NSF] and the National Institutes of Health [NIH]);

(3) a clear need for nonpublic data;

(4) feasibility; and

(5) from the applicant, a willingness to accept all confidentiality protection and disclosure review requirements, including strict limits placed on how much and how often intermediate output can be taken out of the RDC (e.g., by a graduate student for review by his or her professor) and the requirement that the addition of new investigators to a project requires a de novo review and approval.

Stringent application of these criteria may discourage applications or the withdrawal of proposals before a decision is reached and probably contributes to the length of the review process. Yet such stringency arguably does not enhance confidentiality protection nor forward the mission of the Census Bureau to facilitate data use. In particular, the panel concludes that the first criterion has been interpreted in a way that actually impedes furtherance of the agency's mission. Research that uses Title 13 data should be deemed eligible for approval so long as the researcher agrees to provide information to the Census Bureau about the quality and usefulness of the data, without the requirement to demonstrate that the

---

[2]For a description of the Census Bureau's RDC operations, see Hildreth (2003); see also the summary of the panel's workshop, Appendix A.

research's predominant purpose is for data improvement under Title 13 (see Hildreth, 2003; see also discussion in Appendix A). The research use of the data is a key part of the all-important feedback cycle that contributes to improvement of published statistics, public-use microdata, and summary products, as well as the Census Bureau's knowledge about its data.

With regard to the length of review, (incomplete) data from the Census Bureau's RDC network indicate that it takes an average of 7 months for approval of a project with economic data and an average of as much as 20 months for review of a project using matched administrative data, such as state Medicaid and unemployment records matched with CPS or SIPP data. (These figures do not include proposals that require revision and resubmission or that are withdrawn from consideration.) One reason that the use of matched data takes so long to approve is that the custodian of the administrative data undertakes its review following that of the Census Bureau. If a researcher is then asked to revise and resubmit the proposal, and the researcher has, say, a 2-year grant period, the project is impossible to carry out in the allotted time.

Until recently, the process was further slowed because there were only three review cycles each year. In 2004, the Census Bureau implemented a continuous review process in which reviews are being conducted on an "on demand basis." Although it is too early to assess the effects of this change, it is an important step in improving access to confidential data under the RDC program. However, the Internal Revenue Service, because of staff limitations, continues to have only three review cycles each year for projects that propose to use data from Social Security earnings records or income tax returns.[3]

As noted in Chapter 2, the Census Bureau has indicated openness to other ideas for streamlining the application and review process for RDC projects, and some ideas may also be relevant for RDC operations at other agencies. For example, some research projects that are proposed for implementation at an RDC have already been reviewed and recommended for funding by an agency (e.g., the NSF or NIH) through its own peer review process. The Census Bureau (or other agency) could accept that funding recommendation as part of its review process, concentrating only on the appropriateness of the RDC for the work.

---

[3]For a description of the agreement between the Census Bureau and the IRS regarding access to confidential IRS data, see "Criteria for the Review and Approval of Census Projects that Use Federal Tax Information" (www.ces.census.gov/download.php? document=50 [May 2005]). This document applies with full rigor only to proposals using both Title 13 (Census) and Title 26 (IRS) data. When the proposal uses only Title 13 data, the Census Bureau may interpret the statute without IRS review.

Recommendation 9 To achieve the research potential and cost-effective operation of the Census Bureau data centers, the Census Bureau should (1) broaden the interpretation of the criteria for assessing the benefits of access to data; (2) maintain the continuous review cycle; and (3) take account of prior scientific review of research proposals by established peer review processes.

If IRS data are involved, that agency would also have to agree on the new criteria.

Other steps to consider for stimulating use of research data centers include broader advertising of the centers and the procedures for using them, and special proposal submission and review processes for junior researchers. In addition, the Census Bureau and other statistical agencies should explore ways to house confidential data from as many agencies as possible in a single supervised location in a number of host institutions in order to add to their value for research use. The 2002 Confidential Information Protection and Statistical Efficiency Act (CIPSEA) may facilitate this process.

Currently, statistical agencies have few resources to facilitate access to confidential files by external researchers, which is a disincentive to maintain, let alone expand, the operations of research data centers and other modes of restricted access. Similarly, potential host organizations often lack adequate resources to contribute to the operations of research data centers, and they are unlikely to increase their contributions if the access process is so cumbersome that it deters researchers from seeking to use confidential data. In order to provide adequate access, research data centers need funds for a range of tasks, from processing applications and overseeing access, to preparing and updating user-friendly documentation and access tools, to checking researchers' work to ensure that breaches of confidentiality have not occurred. We note that the Census Bureau RDCs are supported in part through grants from the NSF and the National Institute on Aging (see National Research Council, 2000:48). Increased funding through a variety of mechanisms, and from a variety of agencies, should be explored, contingent on improved data access.

EXPANDING AND IMPROVING REMOTE ACCESS

One way to reduce the costs in time and money involved in traveling to a research data center is to expand access to the confidential data stored in those centers from a remote computer. Because the methodology used by an agency in processing and archiving its data affects how remote access to the data can be structured, there are no simple designs for remote access (see Rowland, 2003). Furthermore, access from a remote computer poses significant challenges to the maintenance of confidentiality because of the risk posed by repeated queries to the database and the potential ability to infer individual attributes by comparing results for some table cells against others (see Duncan and Mukherjee, 2000). At this stage of software development for disclosure review, manual monitoring before output is sent back to a user may be more effective at protecting confidentiality. It may also, as in the NCHS system, allow users to request a broader array of outputs (e.g., regressions of various types in addition to tables). However, manual monitoring is more costly for the sponsor agency and precludes rapid response to user submissions.

Research that will permit expansion of this mode of access to confidential data is needed. The research should focus on efficient disclosure limitation methods for remote access that allow users to request a wide range of outputs and obtain output within reasonable time limits.

Recommendation 10 Statistical agencies and other agencies that sponsor data collection should conduct or sponsor research on cost-effective means of providing secure access to confidential data by means of a remote access mechanism, consistent with their confidentiality assurance protocols.

LICENSING AGREEMENTS

An alternative to research data centers, one that reduces burden to users because it does not require them to travel to a different location, is a licensing agreement. Licensing agreements, which are a valuable means of access to confidential data, have developed in different ways for different datasets. Although the Census Bureau does not currently have the authority to allow access to its confidential data under licensing agreements, the Bureau of Labor Statistics, the NCES, and NSF's Division of Science Resources Statistics, among other agencies, license the use of confidential data to researchers who meet certain criteria. The HRS, which is carried out at the University of Michigan with funding from the National Institute on Aging, also licenses researchers to use its data. These licenses enable researchers to work at their home institution, without incurring the costs of relocating.

NCES—which currently uses licensing more than any other agency—requires potential users (such as state and local agencies, contractors, researchers) to complete an application designed for the specific type of user. The process involves preparing and submitting a formal letter of request, a license document, an affidavit of nondisclosure, and a security plan.[4] Users of confidential HRS data must be affiliated with an institu-

---

[4]For details, see "Restricted-Use Data Procedures Manual" (nces.ed.gov/statprog/rudman/[November 2004]).

tion that has a human subjects review process (including an IRB that is registered with and has been approved by the Office for Human Research Protections in the U.S. Department of Health and Human Services) and be a current recipient of federal research funds. Users of HRS data are required to submit for approval a research proposal and a data protection plan, as well as IRB review and a signed agreement for use of restricted data.[5] Most licensing agreements are time limited and require users to return or destroy the confidential data files.

Although potentially very useful for expanding access to confidential data, licensing is not yet widely used by statistical agencies. This mechanism could be significantly expanded: agencies that currently lack authority for licensing should investigate obtaining such authority, and agencies that currently license only a few data sets should consider expanding the number of data sets for which a license may be obtained. In expanding the use of licensing agreements, agencies should sponsor consultations among data users and producers in developing the standards governing such agreements in order to assure the widest possible access consistent with confidentiality protection. Implementation of CIPSEA will facilitate—indeed, require—developing relatively uniform procedures across agencies.

Recommendation 11 Statistical and other agencies that provide data for research and do not yet use licensing agreements for access to confidential data should implement such an access mechanism. Agencies that use licensing for only a few confidential data sets should expand the files for which a license may be obtained.

For some agencies, such a mechanism may require new legislation.

Recommendation 12 Statistical and other agencies that provide data for research should work with data users to develop flexible, consistent standards for licensing agreements and implementation procedures for access to confidential data.

Both the NCES and the HRS licensing agreements include two important enforcement provisions. One is random auditing of the licensed research site by a qualified auditor for adherence to the conditions of the license, including storage of the data on secure servers, restriction of access to personnel named on the agreement, and encryption of the data when in transit. The second is severe penalties for serious violations of the agreement. In the case of the HRS, for example, the penalties include forfeiture by the investigator—and, possibly, the investigator's entire insti-

---

[5]For details, see "HRS Restricted Data: Application Materials: Basic Requirements" (hrsonline.isr.umich.edu/rda/rdapkg_req.htm#reqoutline [November 2004]).

tution—of all current funding, and denial of future funding by the sponsoring agency.

An early review of the results of audits by NCES revealed that the violations uncovered resulted from simple carelessness, did not result in confidentiality breaches, and did not trigger the imposition of penalties (see McMillen, 1999). A more recent review concluded that enforcement mechanisms throughout the government are quite weak (see Seastrom, Wright, and Melnicki, 2003), contributing to violations; however, most if not all of the violations resulted from carelessness or not following proper procedures, rather than from willful misuse of data, and, again, there was no evidence of disclosure of individual data.

Although the panel recognizes that broadening access through licensing agreements may increase the risk of disclosure by increasing the number of people with access to confidential data, we believe that the risk is outweighed by the benefits of wider access. In order to provide as much protection as possible, we recommend that future licensing agreements include the two key enforcement features—auditing and penalties for violations—that are designed to minimize that risk. We also recommend that all data providers be informed that their data may be used in unanticipated ways, and by researchers other than those carrying out the data collection, but only for research purposes (see Recommendation 14).

Recommendation 13 Licensing agreements should include auditing procedures and appropriate legal penalties for willful misuse of confidential data.

INFORMING RESPONDENTS OF DATA USE

As we stress throughout this report, the foundation for achieving the benefits of data for research and policy is the public's willingness to supply the information requested. In turn, all agencies that collect data have an obligation to inform respondents about the purposes for which the data are being collected and how they will be used.

Recommendation 14 Basic information about confidentiality and data access given to everyone asked to participate in statistical surveys should include notification about:

(1) planned record linkages for research purposes;
(2) the possibility of future uses of the data for other research purposes;
(3) the possibility of future uses of the data by researchers other than those collecting the data;
(4) planned nonstatistical uses of the data; and

(5) a clear statement of the level of confidentiality protection that can be legally and technically assured, recognizing that a zero risk of disclosure is not possible.

This recommendation substantially mirrors one from *Private Lives and Public Policies* (National Research Council, 1993:220-221).

The following paragraph may provide a model for a brief statement that responds to the spirit of items (1) – (5).[6]

> Your information is being collected for research purposes and for statistical analysis by researchers in our agency and in other institutions. Your data will not be used for any legal or enforcement purpose [unless required by the Patriot Act]. The researchers who have access to your data are pledged to protect its confidentiality and are subject to fines and prison terms if they violate it. Data will only be provided to researchers outside our agency in a form that protects your identity as an individual. Some uses of your data may require linking your responses to other records, always in a manner that honors our pledge to protect your confidentiality.

The panel also believes that in formulating policies about data access, neither statistical agencies nor IRBs should assume that they know what kinds of data members of the public consider sensitive or what disclosure risks they are willing to tolerate. Instead, these policies should take the views of the public into account.

Recommendation 15 Statistical and funding agencies should support continuing research to monitor the views of data providers and the general public about research risks and benefits, including such topics as the sensitivity of questions, data sharing for statistical purposes, methods of obtaining consent for survey participation, the importance of privacy and confidentiality, and similar topics.

SAFEGUARDING CONFIDENTIALITY: TRAINING, MONITORING, AND EDUCATION

So far, we have discussed ways of expanding research access while protecting confidentiality, focusing mainly on risks of statistical disclosure and how to measure and safeguard against them. In this concluding section, we address the issue of confidentiality protection more generally,

---

[6]Guidance from the Office of Management and Budget on the wording of a confidentiality pledge to be given to respondents participating in data collections subject to CIPSEA is currently in draft form for comment.

acknowledging, as we did in Chapter 4, that wider access to confidential data is likely to increase the risk of confidentiality breaches, but that statistical disclosure is not the only, or even the main, threat. We consider three aspects of confidentiality protection: (1) training employees in procedures to safeguard confidential data, (2) research on violations of confidentiality protection procedures and actual breaches of confidentiality, and (3) educating researchers and staff in the ethical foundations of privacy and confidentiality.

Training Employees

One common threat to confidentiality protection of research data arises from simple carelessness—not removing identifiers from questionnaires or electronic data files, leaving cabinets unlocked, not encrypting files containing identifiers, talking about specific respondents with others not authorized to have this information. Just as institutional review boards currently require researchers to undergo training in human subjects protection issues before undertaking research involving human participants, so statistical agencies and private survey organizations should provide their employees with guidelines for confidentiality protection, as well as regularly updated training in appropriate data management (such as secure storage of identifiable information) to ensure that the guidelines are observed. Data collection agencies that have such guidelines and training should regularly review their procedures to ensure that they are up to date and systematically enforced.

> **Recommendation 16** Statistical agencies and survey organizations that collect individually identifiable data should provide written guidelines for confidentiality protection, as well as training in confidentiality practices and data management that guard against disclosure, for all staff who work with or have access to such data.

Such training should include all aspects of data management—entering, storing, manipulating, and analyzing electronic records. Everyone who handles electronic records needs to be fully aware of the need to protect them, as they do with paper records.

Research on Breaches of Confidentiality

Just as better information is needed about the use made of research data (see Chapter 3), information is also needed about violations of confidentiality protection practices and the actual occurrence of confidentiality breaches. Without knowing how many breaches occur in an agency, it is impossible to know, for example, whether laws and penalties designed to prevent improper disclosure of confidential information are effective or whether other kinds of deterrents are needed. Statistical agencies and individual researchers have generally resisted suggestions for research on confidentiality breaches, yet such research is necessary to evaluate the effectiveness of data access mechanisms in preventing unwarranted disclosure.

The Office of Research Integrity in the U.S. Department of Health and Human Services is currently funding research into such violations of research ethics as data fabrication and plagiarism; a few such studies have been published (see, e.g., Swazey, Anderson, and Louis, 1993; Martinson, Anderson, and de Vries, 2005). Research into the extent, nature, and causes of confidentiality breaches is long overdue. If well-designed and executed research and monitoring finds little evidence of such breaches, it would do much to reassure the public and the agencies themselves that the benefits accruing from wider dissemination of research data will not incur undue costs in terms of breaches of confidentiality. The U.S. Government Accountability Office has in the past expressed an interest in undertaking such research.

> **Recommendation 17** Statistical agencies should set up procedures for monitoring, on an ongoing basis, violations of confidentiality protection practices and instances of confidentiality breaches that may occur. The system should be designed to obtain information on the causes and consequences of these breaches.

Education in Research Ethics

Laws and procedures designed to prevent confidentiality breaches and punish their occurrence will not be optimally effective unless they are accompanied by internalized norms of research ethics and fair information practices (see Barquin and Northouse, 2003; Duncan, 2004). To inculcate these among current and future researchers and the staffs of the statistical agencies, universities as well as agencies that collect data from the public should be encouraged to develop curricula (presented in courses, workshops, and other educational forums) dealing explicitly with the requirements of fair information practices, as well as with the requirements for conducting ethical research with human beings. The two are not identical, and both have a role to play in the training of researchers and others who will work in the field of government statistics. Such education programs should deal with ethical, legal, and data quality issues, as well as with administrative and technical procedures for confidentiality protection, data security, disclosure limitation, and informed consent.

Statistical agencies could make important contributions to the development of such training programs by providing advice based on their experience and expertise. Funding organizations, such as the NSF and the NIH, could contribute to the necessary financial support. There is also an important role in education for professional associations, many of which have codes of professional conduct and ethical standards. Such associations as the American Statistical Association, the American Sociological Association, the Population Association of America, the American Economic Association, the American Association for Public Opinion Research, and their counterparts for other disciplines and fields can contribute significantly to the development of strong norms for fair and ethical practices in research and information gathering.

Recommendation 18 Training in ethical issues related to research, including fair information practices, as well as principles and practices related to research with human participants, should be part of the professional training of all those involved in the design, collection, distribution, and use of data obtained under pledges of confidentiality. Such training should be updated at intervals after the end of formal schooling.

Recommendation 19 Professional associations should develop strong codes of ethical conduct that reflect the need to protect the confidentiality of personal data and make adherence to these codes an integral part of their educational activities.

In addition to encouraging educational and research institutions to add training to their programs, consideration should also be given to requiring completion of a specialized training program as a condition for use of confidential data. Such a program might be designed along the lines of the training and certification programs required of all researchers who are subject to IRBs. Professional associations may be one kind of organization to provide such training.

The challenge facing statistical and other data collection agencies in disseminating the best data as widely as possible in order to foster sound public policy and research while protecting the confidentiality of those data is formidable, but it can be met. With appropriate safeguards, and recognizing that the technological and legal environment is likely to be one of continual change, the nation can reap enormous benefits from the information the public provides.

References

Abowd, J.M., and J.I. Lane
    2003   The Economics of Data Confidentiality. Unpublished paper presented at the National Research Council's Committee on National Statistics Workshop on Confidentiality of and Access to Research Data Files, October 16-17, Washington, DC. Available: http://instruct1.cit.cornell.edu/~jma7/abowd_lane_CNSTAT_economics_ of_confidentiality_2003016.pdf [July 2005].
    2004   New approaches to confidentiality protection: Synthetic data, remote access and research data centers. Pages 282-289 in J. Domingo-Ferrar and V. Torra, editors., *Privacy in Statistical Databases*. New York: Springer-Verlag.

Abowd, J.M., and S.D. Woodcock
    2001   Disclosure limitation in longitudinal linked data. In *Confidentiality, Disclosure and Data Access: Theory and Practical Application for Statistical Agencies*, P. Doyle, J.I. Lane, J.M. Theeuwes, and L.V. Zayatz, editors. Amsterdam, The Netherlands: North-Holland.

American Association of State Colleges and Universities
    2003   *Access for All? Debating In-State Tuition for Undocumented Alien Students*. Available: http://www.aascu.org/special_report/access_for_all.htm [December 3, 2004].

Bailar, J.C., III
    2003   The Role of Data Access in Scientific Replication. Unpublished paper presented at the National Research Council's Committee on National Statistics Confidential Data Access for Research Purposes Workshop, October 16-17, Washington, DC.

Barabba, V.
    1975   The right of privacy and the need to know. In *The Census Bureau: A Numerator and Denominator for Measuring Change*. U.S. Census Bureau Technical Paper 37. Washington, DC: U.S. Government Printing Office.

Barquin, R., and C. Northouse
    2003   Data Collection and Analysis: Balancing Individual Rights and Societal Benefits. Unpublished paper presented at the National Research Council's Committee on National Statistics Confidential Data Access for Research Purposes Workshop, October 16-17, Washington, DC.

Bates, N.A.
    2005   Development and Testing of Informed Consent Questions to Link Survey Data with Administrative Records. *Proceedings of ASA Survey Research Methods Section.* Alexandria, VA: American Statistical Association.
Battacharya, J., and D. Lakdawalla
    2003   *Does Medicare Benefit the Poor? New Answers to an Old Question.* NBER Working Paper No. 9215 revised. Cambridge, MA: National Bureau of Economic Research.
Boisjoly, J., K.M. Harris, and G.T. Duncan
    1998   Trends, events, and duration of initial welfare spells. *Social Service Review* 72(4):466-492.
Bound, J., C. Brown, and N. Mathiowetz
    2001   Measurement error in survey data. In *Handbook of Econometrics*, Volume 5, J. Heckman and E. Leamer, editors. Amsterdam, The Netherlands: North Holland.
Brown, C.
    2003   Longitudinal Data and Public Policy. Unpublished paper presented at the National Research Council's Committee on National Statistics Confidential Data Access for Research Purposes Workshop, October 16-17, Washington, DC.
Cain, G., and D. Wissoker
    1990   A reanalysis of marital stability in the Seattle-Denver Income Maintenance Experiment. *American Journal of Sociology* 95(5):1235-1269.
Clemetson, L.
    2004   Homeland security given data on Arab-Americans. *New York Times*, July 30.
Cohen, S.H., and W. Hadden
    2004   *Issues and Impediments to Expanding Access to Confidential Statistical Agency Data: Restricted Data and Restricted Access.* Statistical Policy Working Paper No. 35, Federal Committee on Statistical Methodology Seminar. Available: http://www.fcsm.gov [September 2005].
Consortium of Social Science Associations
    2004   *Behavioral and Social Science in the Administration's FY 2004 Budget.* Available: www.aaas.org/spp/rd/04pch21.pdf [November 29, 2004].
Couper, M.P., E. Singer, and R.A. Kulka
    1998   Participation in the 1990 decennial census: Politics, privacy, pressures. *American Politics Quarterly* 26: 59-80.
Dalenius, T., and S.P. Reiss
    1982   Data-swapping: A technique for disclosure control. *Journal of Statistical Planning and Inference* (6):73-85.
Dash, E., and T. Zeller
    2005   MasterCard says 40 million files are put at risk. *New York Times* June 18:A1
David, P.A.
    2001   Digital Technologies, Research Collaborations and the Extension of Protection for Intellectual Property in Science: Will Building 'Good Fences' Really Make 'Good Neighbors' in Science? MERIT Research Memorandum No. 2001-004. Maastrict Economic Research Institute on Innovation and Technology, Maastrict, The Netherlands.
Dillman, D.A., M.D. Sinclair, and J.R. Clark
    1993   Effects of questionnaire length, respondent-friendly design, and a difficult question on response rates for occupant-addressed census mail surveys. *Public Opinion Quarterly* 57:289-304.
Domingo-Ferrer, J., and V. Terra
    2001   Disclosure control methods and information loss for microdata. In *Confidentiality, Disclosure and Data Access: Theory and Practical Application for Statistical Agencies*, P. Doyle, J.I. Lane, J.M. Theeuwes, and L.V. Zayatz, editors. Amsterdam, The Netherlands: North-Holland.

Doyle, P., J.I. Lane, J.J.M., Theeuwes, and L.V. Zayatz.
2001 *Confidentiality, Disclosure, and Data Access: Theory and Practical Applications for Statistical Agencies.* Amsterdam, The Netherlands: North-Holland.

Duncan, G.T.
2002 Confidentiality and statistical disclosure limitation. In *International Encyclopedia of the Social and Behavioral Sciences,* N.J. Smelser and P.B. Baltes, editors. Oxford, U.K.: Pergamon.
2004 Exploring the tension between privacy and the social benefits of governmental databases. In *A Little Knowledge: Privacy, Security and Public Information After September 11,* P.M. Shane, J. Podesta, and R.C. Leone editors. New York: The Century Foundation.

Duncan, G.T., S.E. Fienberg, R. Krishnan, R. Padman, and S.F. Roehrig
2001 Disclosure limitation methods and information loss for tabular data. In *Confidentiality, Disclosure and Data Access: Theory and Practical Applications for Statistical Agencies,* P. Doyle, J.I. Lane, J.M. Theeuwes, and L.V. Zayatz, editors. Amsterdam, The Netherlands: North-Holland.

Duncan, G.T., S. Keller-McNulty, and S.L. Stokes
2003 *Disclosure Risk vs. Data Utility: The R-U Confidentiality Map.* Working Paper. Available: http://www.heinz.cmu.edu/wpapers/detail.jsp?id=4386 [November 29, 2004].

Duncan, G.T., and D. Lambert
1986 Disclosure-limited data dissemination. *Journal of the American Statistical Association* 81:10-18.
1989 The risk of disclosure for microdata. *Journal of Business and Economic Statistics* 7:207-217.

Duncan, G.T., and S. Mukherjee
2000 Optimal disclosure limitation strategy in statistical databases: Deterring tracker attacks through additive noise. *Journal of the American Statistical Association* (95):720-729.

Duncan G.T., and R.W. Pearson
1991 Enhancing access to microdata while protecting confidentiality. *Statistical Science* 6:219-239.

Duncan, G.T., and S.L. Stokes
2004 Disclosure risk vs. data utility: The R-U confidentiality map as applied to topcoding. *Chance* 17:16-20.

Duncan, K.
2000 Incentives and the work decisions of welfare recipients: Evidence from the Panel Survey of Income Dynamics, 1981-1988. *American Journal of Economics and Sociology* 59(3):433-449.

Dunton, N.
2000 PUMS. Pp. 311-312 in *Encyclopedia of the U.S. Census,* M.J. Anderson, editor. Washington, DC: CQ Press.

Ferber, R.
1966 *The Reliability of Consumer Reports of Financial Assets and Debt.* Studies in Consumer Savings # 6. Urbana, IL: Bureau of Economic and Business Research, University of Illinois.

Foster, L., J. Haltiwanger, and C.J. Krizan
2001 Aggregate productivity growth: Lessons from microeconomic evidence. Chapter 8 in *New Directions in Productivity Analysis,* E. Dean, M. Harper, and C. Hulten, editors. Chicago: University of Chicago Press.

Fox, S., and O. Lewis
- 2001 *Fear of Online Crime: Americans Support FBI Interception of Criminal Suspects' Email and New Laws to Protect Online Privacy.* Pew Internet & American Life Project. Available: http://www.pewinternet.org.

Gaquin, D.
- 2000a Data dissemination and use. In *Encyclopedia of the U.S. Census*, M.J. Anderson, editor in chief. Washington, DC: CQ Press.
- 2000b Summary tape files. *Encyclopedia of the U.S. Census*, M.J. Anderson, editor. Washington, DC: CQ Press.

Garber, A.M., T.E. MaCurdy, and M.C. McClellan
- 1998 Persistence of Medicare expenditures among elderly beneficiaries. Pp. 153-180 in *Frontiers in Health Policy*, Vol. 1, A.M. Garber, editor. Cambridge, MA: MIT Press.

Gilheany, S.
- 2000 *The Decline of Magnetic Disk Storage Cost Over the Next 25 Years.* Available: http://www.berghell.com/whitepapers/Storage%20Costs.pdf [December 2, 2004].

Gordon, R.
- 1999 Confidential data files linked to the National Longitudinal Survey of Youth, 1979 Cohort: A case study. Unpublished paper presented at the National Research Council's Committee on National Statistics Workshop on Confidentiality of and Access to Data Research Files, October 14-15, Washington, DC.

Groenveld, L. P., N.B. Tuma, and M.T. Hannan
- 1980 The effects of negative income tax programs on marital dissolution. *The Journal of Human Resources* 15(4)(fall):654-674.

Groves, R., and J. Lepkowski
- 1985 Cost and error modeling for large-scale telephone surveys. Pp. 330-357 in *Proceedings of the Bureau of the Census First Annual Research Conference.* Washington, DC: U.S. Department of Commerce.

Guarino, J.A., J.M. Hill, and H.F. Woltman
- 2001 *Analysis of the Social Security Number - Notification Component of the Social Security Number, Privacy Attitudes, and Notification Experiment.* Washington, DC: U.S. Department of Commerce.

Haltiwanger, J., S. Davis, and S. Schuh
- 1996 *Job Creation and Destruction.* Cambridge, MA: MIT Press.

Hawala, S., L. Zayatz, and S. Rowland
- 2004 American FactFinder: Disclosure limitation for the advanced query system. *Journal of Official Statistics* 20(1):115-124.

Hayes, B.
- 2002 Terabyte territory. *American Scientist* 90(3):212-216.

Hildreth, A.K.
- 2003 The Census Research Data Center (RDC) Network: Problems, Possibilities and Precedents. Unpublished paper presented at the National Research Council's Committee on National Statistics Confidential Data Access for Research Purposes Workshop, October 16-17, Washington, DC.

Hillygus, S., N. Nie, and K. Prewitt with G. Pals
- 2006 *Civic Mobilization and Privacy Concerns in the 2000 Census.* New York: Russell Sage Foundation.

Hoynes, H., and T. MaCurdy
- 1994 Has the decline in benefits shortened welfare spells? *American Economic Review* 84(2):43-48.

Hurd, M., F.T. Juster, and J.P. Smith
    2003   Enhancing the quality of data on income: Recent innovations from the HRS. *Journal of Human Resources* 38(3):758-772.

Institute of Medicine
    2005   *Vaccine Safety Research, Data Access, and Public Trust*. Committee on the Review of National Immunization Program's Research Procedures and Data Sharing Program, Board on Health Promotion and Disease Prevention. Washington, DC: The National Academies Press.

Journal of Human Resources
    1995   Vol. 30, No 5. Available: http://www.ssc.wisc.edu/jhr/1995ab/loprest.html [May 23, 2005].

Klein, L., and A.C. Goldberger
    1955   *An Econometric Model of the United States, 1929–1952*. Amsterdam: North-Holland Publishing Co.

Kuhn, T.
    1962   *The Structure of Scientific Revolutions*. Chicago, IL: University of Chicago Press.

Lambert, D.
    1993   Measures of disclosure risk and harm. *Journal of Official Statistics* 9:313-333.

Levitan, S.A., and F. Gallo
    1990   Work and family: The impact of legislation. *Monthly Labor Review*, March, 34-40.

Linet, M.
    2003   Impact of HIPAA on Research. Testimony prepared for the National Committee on Vital and Health Statistics Subcommittee on Privacy and Confidentiality, November 19-20, Silver Spring, MD.

Little, R.J.A.
    1993   Statistical analysis of masked data. *Journal of Official Statistics* 9:407-426.

Marquis, K.H., and C.J. Moore
    1990   Measurement errors in SIPP program reports. Pp. 721-745 in *Proceedings of the Bureau of the Census 1990 Annual Research Conference*. Washington, DC: U.S. Department of Commerce.

Martinson, B.C., M.S. Anderson, and R. de Vries
    2005   Scientists behaving badly. *Nature* 435:737-738.

McClellan, M., and J. Skinner
    2004   *The Incidence of Medicare*. (Working Paper.) Available: http://www.dartmouth.edu/~economic/faculty/Skinner/Papers/ms%20medicare%206%20feb%2004.pdf [November 30, 2004].

McCullagh, D.
    2004   Database nation: The upside of "zero privacy." *Reason* (June). Available: http://www.reason.com/0406/fe.dm.database.shtml [December 21, 2004].

McGuckin, R.H.
    1992   The Importance of Establishment Data in Economic Research. Discussion paper, Center for Economic Studies. U.S. Census Bureau, Washington, DC.
    1995   Establishment microdata for economic research and policy analysis: Looking beyond the aggregates. *Journal of Business and Economic Statistics* 13(1):121-126.

McGuckin, R.H., and S.V. Nguyen
    1990   Public use microdata: Disclosure and usefulness. *Journal of Economic and Social Measurement* 16(1):19-39.

McMillen, D.
    2003   Privacy, Confidentiality, and Data Sharing. Unpublished paper presented at the National Research Council's Committee on National Statistics Confidential Data Access for Research Purposes Workshop, October 16-17, Washington, DC.

McMillen, M.
    1999   National Center for Education Statistics: Data licensing systems. Unpublished paper presented at the National Research Council's Committee on National Statistics Workshop on Confidentiality of and Access to Data Research Files, October 14-15, Washington DC.

Menchik, P.L., and M. David
    1983   Income distribution, lifetime savings, and bequests. *American Economic Review* 73(4):672-690.

National Commission for the Protection of Human Subjects of Biomedical and Behavioral Research
    1979   *The Belmont Report: Ethical Principles and Guidelines for the Proection of Human Subjects of Research.* Office of Human Subjects Research, National Institutes of Health. Washington, DC: U.S. Department of Health and Human Services.

National Highway Traffic Safety Administration
    2001   *National EMS Research Agenda.* Available: http://www.nhtsa.dot.gov/people/injury/ems/ems-agenda/toc.htm [December 6, 2004].

National Research Council
    1979   *Privacy and Confidentiality as Factors in Survey Response.* Panel on Privacy and Confidentiality as Factors in Survey Response, Committee on National Statistics, Asssembly of Behavioral and Social Sciences. Washington, DC: National Academy Press.
    1985   *Sharing Research Data.* Committee on National Statistics. S.E. Fienberg, M.E. Martin, and M.L. Straf, editors. Commission on Behavioral and Social Sciences and Education. Washington, DC: National Academy Press.
    1991   *Improving Information for Social Policy Decisions: The Uses of Microsimulation Modeling. Volume I: Review and Recommendations.* Panel to Evaluate Microsimulation Models for Social Welfare Programs, C.F. Citro and E.A. Hanushek, editors. Committee on National Statistics, Commission on Behavioral and Social Sciences and Education. Washington, DC: National Academy Press.
    1993   *Private Lives and Public Policies: Confidentiality and Accessibility of Government Statistics.* Panel on Confidentiality and Data Access, G.T. Duncan, T.B. Jabine, and V.A de Wolf, editors. Committee on National Statistics, Commission on Behavioral and Social Sciences and Education. Washington, DC: National Academy Press.
    1997   *Assessing Policies for Retirement Income: Needs for Data, Research, and Models.* Panel on Retirement Income Modeling, C.F. Citro and E.A. Hanushek, editors. Committee on National Statistics, Commission on Behavioral and Social Sciences and Education. Washington, DC: National Academy Press.
    1998   *People and Pixels: Linking Remote Sensing and Social Science.* Committee on the Human Dimensions of Global Change, Diana Liverman, Emilio F. Moran, Ronald R. Rindfuss, and Paul C. Stern, editors. Commission on Behavioral and Social Sciences and Education. Washington, DC: National Academy Press.
    2000   *Improving Access to and Confidentiality of Research Data: Report of a Workshop.* Committee on National Statistics, C. Mackie and N. Bradburn, editors. Commission on Behavioral and Social Sciences and Education. Washington, DC: National Academy Press.
    2001a   *Cells and Surveys: Should Biological Measures Be Included in Social Science Research?* Committee on Population, C.E. Finch, J.W. Vaupel, and K. Kinsella, editors. Commission on Behavioral and Social Sciences and Education. Washington, DC: National Academy Press.

2001b *Evaluating Welfare Reform in an Era of Transition.* Panel on Data and Methods for Measuring the Effects of Changes in Social Welfare Programs, R.A.Moffitt and M. Ver Ploeg, editors. Committee on National Statistics, Division of Behavioral and Social Sciences and Education. Washington, DC: National Academy Press.

2002 *Access to Research Data in the 21st Century: An Ongoing Dialogue Among Interested Parties. Report of a Workshop.* Science, Law, and Policy Panel, Policy and Global Affairs. Washington, DC: National Academy Press.

2003a *Ensuring the Quality of Data Disseminated by the Federal Government: Workshop Report.* Ad Hoc Committee on Ensuring the Quality of Government Information, Science, Technology, and Law Program, Policy and Global Affairs. Washington D.C.: The National Academies Press.

2003b *Protecting Participants and Facilitating Behavioral and Social Science Research.* Panel on Institutional Review Boards, Surveys, and Social Science Research, C.F. Citro, D.R. Ilgen, and C.B. Marrett, editors. Committee on National Statistics and Board on Behavioral, Cognitive, and Sensory Sciences, Division of Behavioral and Social Sciences and Education. Washington, DC: The National Academies Press.

2004a *Measuring Racial Discrimination.* Panel on Methods for Assessing Discrimination, R.M. Blank, M. Dabady, and C.F. Citro, editors. Committee on National Statistics, Division of Behavioral and Social Sciences and Education. Washington, DC: The National Academies Press.

2004b *The 2000 Census: Counting Under Adversity.* Panel to Review the 2000 Census, C.F. Citro, D.L. Cork, and J.L. Norwood, editors. Committee on National Statistics, Division of Behavioral and Social Sciences and Education. Washington, DC: The National Academies Press.

2005 *Principles and Practices for a Federal Statistical Agency, Third Edition.* Committee on National Statistics. M.E. Martin, M.L. Straf, and C.F. Citro, editors. Division of Behavioral and Social Sciences and Education. Washington, DC: The National Academies Press.

Ochas, S., J. Rasmussen, C. Robson, and M. Salib
 2001 Reidentification of Individuals in Chicago's Homicide Database: A Technical and Legal Study. Unpublished manuscript, Massachusetts Institute of Technology.

O'Neill, J., D. Wolf, L. Bassi, and M. Hannan
 1984 *An Analysis of Time on Welfare.* Washington, DC: The Urban Institute.

O'Rourke, J.M., and M.P. Gutmann
 2005 Citations Database—Human Subjects Protection and Disclose Risk Analysis, Project 3: Statistical Disclosure Control: Best Practices and Tools for the Social Sciences. Inter-university Consortium for Political and Social Research, Ann Arbor, MI. Available: www.icpsr.umic.edu/HSP/bibliography-keyword.htm. [October 2005].

Paas, G.
 1988 Disclosure risk and disclosure avoidance for microdata. *Journal of Business and Economic Statistics* 6:487-500.

Perrit, H.H., Jr.
 2003 Protecting Confidentiality of Research Data through Law. Unpublished paper presented at the National Research Council's Committee on National Statistics Confidential Data Access for Research Purposes Workshop, October 16-17, Washington, DC.

Prewitt, K.
 2000 Remarks prepared for session on Census 2000 organized by the Census 2000 Initiative and the Leadership Conference on Civil Rights, March 24, Washington DC.
 2004 What if we give a census and no one comes? *Science* (676):1452-1453.

Raghunathan, T.E.
    2003   Evaluation of Inferences from Multiple Synthetic Data Sets Created Using Semiparametric Approach. Unpublished paper presented at the National Research Council's Committee on National Statistics Confidential Data Access for Research Purposes Workshop, October 16-17, Washington, DC.

Raghunathan, T.E., J.P. Reiter, and D.R. Rubin
    2003   Multiple imputation for statistical disclosure limitation. *Journal of Official Statistics* 19:1-16.

Reiter, J.P.
    2003   Estimating Probabilities of Identification for Microdata. Unpublished paper presented at the National Research Council's Committee on National Statistics Confidential Data Access for Research Purposes Workshop, October 16-17, Washington, DC. [Submitted to *Journal of the American Statistical Association*.]

Rhea, S., P. Eaton, D. Geels, H. Weatherspoon, B. Zhao, and J. Kubiatowicz
    2003   Pond: The OceanStore prototype. *Proceedings of the 2nd USENIX Conference on File and Storage Technologies* (FAST 03). Available: http://www.cs.ucsb.edu/~ebelding/courses/276/f04/papers/pond.pdf [August 2005].

Rindfuss, R.R.
    2002   Conflicting demands: Confidentiality promises and data availability. *Newsletter of the International Human Dimensions Programme on Global Environmental Change*, February. Available: http://www.ihdp.uni-bonn.de/html/publications/update/update02_02/Update02_02_art1.html [November 30, 2004].

Rowland, S.
    2003   Examination of Monitored, Remote Microdata Access Systems. Paper presented at the National Research Council's Committee on National Statistics Confidential Data Access for Research Purposes Workshop, October 16-17, Washington, DC.

Rubin, D.B.
    1993   Discussion of statistical disclosure limitation. *Journal of Official Statistics* (9):461-468.

Ruggles, S.
    2000   Data user's perspective on confidentiality. *Of Significance—Journal of the Association of Public Data Users* 2:1-5.

Schweinhart, L.J.
    2004   *The High/Scope Perry Preschool Study Through Age 40*. Available: www.highscope.org/Research/PerryProject/PerryAge40SumWeb.pdf [December 3, 2004].

Seastrom, M.
    2001   Licensing. Pp. 279-296 in *Confidentiality, Disclosure, and Data Access: Theory and Practical Applications for Statistical Agencies*, P. Doyle, J.I. Lane, J.J.M. Theeuwes, and L.V. Zayatz, editors. Amsterdam, Netherlands: North-Holland, Elsevier Science.

Seastrom, M.M., C. Wright, and J. Melnicki
    2003   The Role of Licensing and Enforcement Mechanisms in Promoting Access and Protecting Confidentiality. Unpublished paper presented at the National Research Council's Committee on National Statistics Confidential Data Access for Research Purposes Workshop, October 16-17, Washington, DC

Seltzer, W., and M. Anderson
    2000   After Pearl Harbor: The Proper Role of Population Data Systems in Time of War. Available: http://www.uwm.edu/~margo/govstat/integrity.htm [December 15, 2004].

2003 Government Statistics and Individual Safety: Revisiting the Historical Record of Disclosure, Harm, and Risk. Paper presented at the National Research Council's Committee on National Statistics Confidential Data Access for Research Purposes Workshop, October 16-17, Washington, DC. Available: http://www.uwm.edu/~margo/govstat/WS-MAcnstat.pdf [December 8, 2004].

Sieber, J.
1991 *Sharing Social Science Data: Advantages and Challenges.* Newbury Park, CA: Sage Publications.

Singer, E.
2004 Principles and practices related to scientific integrity. In *Survey Methodology*, R.M. Groves et al. New York: Wiley.

Singer, E., N.A. Mathiowetz, and M.P. Couper
1993 The impact of privacy and confidentiality concerns on survey participation: The case of the 1990 census. *Public Opinion Quarterly* 57(4):465-482.

Singer, E., R.Y. Shapiro, and L.R. Jacobs
1997 Privacy of health care data: What does the public know? How much do they care? In *Health Care and Information Ethics*, A.y Chapman, ed. Washington, DC: American Association for the Advancement of Science.

Singer, E., J. Van Hoewyk, and R. Neugebauer
2003 Attitudes and behavior: The impact of privacy and confidentiality concerns on participation in the 2000 census. *Public Opinion Quarterly* 65(3):368-384.

Singer, E., J. Van Hoewyk, and R. Tourangeau
2001 Final Report on the 1999-2000 Surveys of Privacy Attitudes. Contract No. 50-YABC-7-66019, Task Order No. 46-YAB-9-0002. Washington, DC: U.S. Census Bureau.

Singh, A.C., F. Yu, and G.H. Dunteman
2003 MASSC: A New Data Mask for Limiting Statistical Information Loss and Disclosure. Paper presented at the Joint ECE/EUROSTAT Work Session on Data Confidentiality, Luxembourg. Available: Statistics Research Division, RTI International, Research Triangle Park, NC.

Soete, L., and Bas ter Weel
2003 ICT and Access to Research Data: An Economic Review. Unpublished paper (June). Maastrict Economic Research Institute on Innovation and Technology, Maastrict, The Netherlands.

Spruill, N.L.
1983 The confidentiality and analytic usefulness of masked business microdata. In *Proceedings of the Section on Survey Research Methods*, American Statistical Association. Washington, DC: American Statistical Association.

Swazey, J., M.S. Anderson, and K. S. Louis
1993 Ethical problems in academic research. *American Scientist Online* (originally appeared in November/December issue). Available: http://www.americanscientist.org/template/AssetDetail/assetid/33432;jsessionid=baagLvqUb8uc7F?fulltext=true [May 31, 2005].

Sweeney, L.
1999 Computational disclosure control for medical microdata: The datafly system. In *Record Linkage Techniques 1997: Proceedings of an International Workshop and Exposition*. Committee on Applied and Theoretical Statistics. Washington, DC: National Academy Press.
2001 Information explosion. In *Confidentiality, Disclosure, and Data Access: Theory and Practical Applications for Statistical Agencies*, P. Doyle, J.I. Lane, J.M. Theeuwes, and L.V. Zayatz, editors. Washington, DC: Urban Institute.

Sweeney, L., B. Malin, and E. Newton
 2003 Trail Re-identification: Learning Who You are From Where You Have Been. Data Privacy Laboratory Technical Report, LIDAP-WP12. Carnegie Mellon University, School of Computer Science.

Taueber, C.
 1981 America enters the eighties: Some social indicators. *The Annals of the American Academy of Political and Social Science* (vol. 453). Thousand Oaks, CA: Sage Publications.

Triplett, J.
 1991 The federal statistical system's response to emerging data needs. *Journal of Economic and Social Measurement* 17(3-4):155-177.

U.S. Census Bureau
 2004 *U.S. Census Bureau Privacy Principles*. Washington, DC: U.S. Department of Commerce.

U.S. Office of Management and Budget
 2004 *Statistical Programs of the United States Government: Fiscal Year 2005*. Washington, DC: Executive Office of the President. Also available: http://www.whitehouse.gov/omb/inforeg/05statprog.pdf [September 2005].
 2005 *Final Information Quality Bulletin for Peer Review*. 70 FR 2664-02 (January 14).

Willenborg, L., and T. de Waal
 1995 *Statistical Disclosure Control in Practice*. (Lecture Notes in Statistics 111.) New York: Springer.

Winkler, W.E.
 1988 Using the EM algorithm for weight computation in the Fellegi-Sunter model of record linkage. Pp. 667-671 in *Proceedings of the Section on Survey Research Methods*. Washington, DC: American Statistical Association.
 1998 Re-identification methods for evaluating the confidentiality of analytically valid microdata. *Research in Official Statistics* 1:87-104.

Woodbury, R., A. Gustman, L. Lillard, O. Mitchell, and R. Willis
 1999 The Value of Linked Data in Aging Research. Paper prepared for April 17-18, 2000, meeting of the Panel on a Research Agenda and New Data for an Aging World, Committee on Population, National Research Council, Washington, DC.

APPENDIX A

Workshop Summary

ACCESS TO RESEARCH DATA: ASSESSING RISKS AND OPPORTUNITIES
OCTOBER 16-17, 2003

The panel held a workshop early in its deliberations to hear from experts about how microdata can best be made available to researchers while protecting respondent confidentiality. The workshop goals were to generate information for the panel's use and to provide a venue for the papers commissioned by the panel to be presented and discussed in a public forum.

This summary, following the workshop, is organized around the six topics that were the subjects of the commissioned papers:

- the changing legal landscape;
- facilitating data access;
- measuring the risks and costs of disclosure;
- the impact of multiple imputation on disclosure risk and information utility;
- assessing the benefits of researcher access to longitudinal microdata; and
- assessing research and policy needs—the economics of data access.

The papers, presenters, and discussants are listed at the end of the

Appendix. The papers are available electronically (www7.national academies.org/cnstat/Data_Access_Panel.html).

BACKGROUND AND OVERVIEW

In 1999 the Committee on National Statistics (CNSTAT) held a workshop focused on the procedures used by agencies and organizations for releasing public-use microdata files and for establishing restricted access to nonpublic files. Tradeoffs between research and other data user needs and confidentiality requirements were articulated, as were the relative advantages and costs of data alteration techniques versus restricted (physical) access arrangements. The report of that workshop, *Improving Access to and Confidentiality of Research Data* (National Research Council 2000), provided a starting point for the panel's work.

The panel's workshop followed up on many of the topics discussed in the 1999 workshop, but the focus was less on what agencies are currently doing and more toward emerging opportunities, specifically relating to research access to longitudinal microdata. Participants provided an indepth look at a number of topics ranging from the role of licensing and penalties for infringing on licensing agreements to the potential of data linking (e.g., between survey and administrative data), particularly the technical, legal, and statistical arrangements that would be needed to promote linking within and between agencies and between government and private-sector data producers. The workshop also sought to promote discussion of how to measure the risks and costs associated with data use, disclosure, and limiting access; what levels of risk are acceptable; and public perceptions about privacy as they pertain to market data in comparison with government data.

The first day was devoted to three topics on risks and opportunities: legal, technical and organizational, and normative. The papers and discussion in the first session examined various aspects of the legal landscape, emphasizing important recent changes, particularly the Confidential Information Protection and Statistical Efficiency Act of 2002 (CIPSEA). The papers highlighted that the legal framework offers a range of opportunities for promoting wider access to research data. For example, through the use of licensing agreements, some of the legal responsibility for maintaining confidentiality can be shifted to data users by the agencies that collect the data.

The papers and discussion in the second session focused primarily on technical and organizational opportunities, both on a general level and as manifested by special organizations like the Census Bureau's Research Data Centers and remote access to data that are stored centrally. The third session was devoted to the difficult problem of accurately assessing dis-

APPENDIX A

closure risks associated with access to microdata and also to the question of what if any harms have come to participants as a result of the disclosure of the information they provided to government agencies.

The second day of the workshop focused on ways of dealing with potentially conflicting goals—information utility and confidentiality protection. Session four focused on one particular method for accommodating those two values, the creation of imputed data, and assessed how well analyses based on such data might approximate models estimated from unaltered data.

Session five was devoted primarily to discussing the scientific and practical benefits of providing restricted as well as unrestricted access to research data. Participants also considered scientific replication and the usefulness of access to data by multiple parties in that process. The final session attempted to assess costs and benefits associated with different approaches to providing data access and protecting confidentiality.

SESSION I:
THE CHANGING LEGAL LANDSCAPE

David McMillen opened the session with the presentation of his paper "Privacy, Confidentiality, and Data Sharing," which reviewed new legislation dealing with these issues. He also provided an overview of the legislative history of CIPSEA, which is Title V of the E-Government Act of 2002, and offered some thoughts about its implementation.

McMillen focused on the principle and application of informed consent agreements. He underscored the point that the legislative history on privacy indicates that when people provide information to the government, or to a private entity, they do not give up all rights to how those data are used. Conversely, when the government receives information from the public, it is not free to use that information for purposes other than those for which the information was collected. The central issue for McMillen, then, is what the terms of this implied contract between data providers and their government are and what responsibility the government has for making those terms clear and explicit.

McMillen argued that agencies that collect information from the public should be as clear and detailed as possible in explaining to respondents how the information will be used and what the limits of confidentiality protection are. He concluded with the statement that government is based on open access to the citizens it serves, and that openness should be one of the principles that guide the development of policies about informing respondents of their rights and responsibilities when asked for information.

During the discussion of McMillen's paper, there was considerable divergence of opinion about how detailed informed consent agreements should or could be. Some participants articulated the view that McMillen's prescriptions would lead to a serious decrease in the utility and value of government-collected microdata.

Marilyn Seastrom, Candice Wright, and John Melnicki then presented their paper, "The Role of Licensing and Enforcement Mechanisms in Promoting Access and Protecting Confidentiality." Licensing agreements allow researchers to use protected confidental data files in a secure environment at their home institution, subject to the terms and responsibilities specified in an agreement. In the first part of their presentation, the authors reviewed the strengths and weaknesses of various licensing arrangements currently in practice in the United States and abroad. They also described instances when enforcement has led to sanctions and when administrative penalties have been implemented for misuse of data. The authors also reviewed application procedures and data security plans and the many types of data agreements in place.

The authors concluded that the enforcement mechanisms are, for a number of reasons, quite weak and that, consequently, there have been violations, though most of them are relatively minor (e.g., computers left unattended, failure to maintain a log for check in/out of data, or data not properly stored when not in use). They concluded with three recommendations: (1) more routine use of security inspections, (2) implementation of termination procedures, and (3) maintenance of a tracking database.

First, given the potential importance of security inspections as a means of monitoring and enforcement of data-use agreements, all agencies that license external researchers to use confidential microdata files should give serious consideration to instituting security inspections on a regular basis. Second, to meet the legal requirements associated with individually identifiable data, entities licensing the use of confidential data must have procedures in place for monitoring the disposition of the data files at the completion of a research project. This requirement can help ensure that the data are not subsequently used for unauthorized purposes. Third, to run an effective data-use agreement program, an agency must have and maintain complete, accurate, and thorough records for each data agreement. Such records are essential for monitoring the authorized users, the approved uses of the data, and the security of the data.

Henry H. Perritt, Jr., concluded the opening session with the presentation of his paper, "Efficacy of Different Theories of Enforcement." The paper suggests a framework within which the efficacy of legal protections of confidentiality can be evaluated, offers qualitative standards for evaluating the effectiveness of existing law, and identifies alternative approaches for strengthening legal protections. The paper begins with an evaluation of the possibility that federal or state law might compel researchers to disclose confidential data received from federal government sources. It identifies the two kinds of private interests that warrant shielding data from disclosure and the sources of law that prohibit disclosure of data identified as confidential by the government agency from which it was received.

Perritt concluded that legal liability is only a weak protection for data confidentiality because the principal privacy statutes do not recognize private rights of action for wrongful disclosure, and the case law under common-law legal theories provides sparse support, at best, for recovery for disclosure. Moreover, difficulties in detection, proof, establishment of damages, and the high cost of litigation make it unlikely that victims of wrongful disclosure would seek relief in the courts. Perritt noted that at least one respected commentator agrees with these shortcomings of existing privacy law.

Perritt proposed two promising ways to afford legal protection against wrongful research disclosure: (1) to require researchers who receive confidential data to establish internal protections, on pain of contract cancellation and bars to receiving grants in the future, and (2) to put nondisclosure language in license agreements that supports "third-party-beneficiary" recovery by data subjects.

During the discussion of Perritt's paper, participants said that an additional protection exists because the institution at which the researcher works has the ability to discipline the researcher further; bringing the institution into the arrangement can strengthen the potential sanctions for disclosure. Perritt concurred, recommending that the design of the institutional mechanism should make clear that the individual researchers have responsibility and accountability and that they will be subject to discipline or discharge if they violate their obligations under the agreement. Furthermore, institutional liability itself is important since, in some sense, the institution has more at risk then does an individual. This incentive can be exploited to promote conformity to data protection rules.

Joe Cecil, one of the formal discussants for the session, said he found Perritt's argument—that it would be difficult to create a meaningful right of private action for an individual and have it work in a way that would give agencies confidence that they are not left responsible for a breach of confidentiality—convincing. He argued that the notion expressed at the 1999 workshop of transferring this responsibility to data users and at the same time giving agencies greater confidence is perhaps a false hope. He suggested that perhaps the focus should be on how to strengthen the institutional mechanisms that Seastrom and others explored and developing data-use agreements for researchers who want to download public-use files.

Katherine Wallman, the second formal discussant, provided extensive clarification about the specifications of CIPSEA. She noted that CIPSEA is the culmination of the work of not just four Congresses, but also almost 25 years of work by her, her predecessors as chief statistician of the U.S. Office of Management and Budget, and many others. The legislation is only the most recent in a long history of efforts to strengthen the legal protection of confidentiality for statistical information collected by the federal government.

The other major objective addressed in CIPSEA concerns the sharing of information among agencies with various kinds of confidential protections and others who are legitimate users of the information, including licensed researchers at universities, licensed researchers in public-sector organizations and pro bono organizations, and others. She noted, however, that the dual objectives of protecting confidentiality and increasing data sharing have sometimes caused confusion about what is in CIPSEA and what is not and about who is covered and who is not covered. For example, although only three named agencies are covered by the data-sharing provisions of the legislation, CIPSEA's confidential protections extend to all federal agencies that collect statistical data under a pledge of confidentiality. Wallman concluded her remarks by briefly outlining the plans for implementing CIPSEA's provisions.

SESSION II:
FACILITATING DATA ACCESS

Michael Larsen gave a presentation on the technical, legal, and organizational barriers to data linkage. Larsen first identified the benefits motivating the goal of data linkage—how such data would be used to enhance analyses. He then discussed technical and legal barriers inhibiting data linkage. He concluded by discussing the role of data enclaves and the example of data linking between the Health and Retirement Study (HRS) and Social Security Administration (SSA) records.

Several themes emerged from the presentation. First, Larsen clearly articulated the importance of data access and data linkages to research, noting examples of questions that could not be answered without access to linked data. In addition to the technical challenges associated with accurately matching records across sources, it is important when seeking respondents' permission at the beginning of a project to think carefully about potential linkages. Proactive work is needed both to make linkage possible and to have respondents' support.

The paper by Andrew Hildreth, "The Census Research Data Center Network: Problems, Possibilities and Precedents," assessed the track record of research data centers (RDCs) and the potential and problems associated with them. The RDC system stems from the desire to permit access to confidential data sets housed at the U.S. Census Bureau. The program started as a pilot in 1994; it was initially funded through the U.S. National Science Foundation (NSF) in partnership with the National Bureau of Economic Research (NBER). The goal was to make such data sets as the longitudinal research database more accessible to researchers by making them available in locations other than Washington, D.C.

Hildreth began with an overview of how to apply for access, the kinds of research projects that are undertaken, and what applicants can expect in terms of process, particularly how long it might take to get to an RDC and start working with the data. He also discussed questions relating to the long-term financial prospects of the data centers. The key problem that Hildreth focused on was that of time delays. He spoke strongly in favor of a system that allows continual review of project applications.

In conclusion, Hildreth said his most important recommendation was to improve the proposal submission and review process for junior users. Wider access will bring wider recognition of what the research has meant to the Census Bureau's data programs and what work the RDCs do and can do. Second, RDCs can be a way to achieve better alignment of the data programs with the Census Bureau's goals. Third, some kind of core funding would be very helpful, perhaps through local institutional support, so that RDCs and the researchers who want to use them do not face yearly worries around budget time.

J. Bradford Jensen related his experiences "from the trenches" in trying to design a national framework for a data enclave model. He characterized the RDC enterprise as expensive, fragile, and tenuous. He suggested that the U.S. Census Bureau experience is representative of those of other countries and other contexts. Jensen confirmed many of Hildreth's observations about the difficulties that lie ahead for the RDC system. However, he, too, noted the immense potential to advance research at the RDCs and was hopeful that the obstacles to their continued and improved operation could be overcome.

Sandra Rowland presented a paper, "An Examination of Monitored, Remote Microdata Access Systems," that focuses on monitored remote (electronic) access to confidential microdata. Many national statistical offices disseminate microdata in three ways: public-use microdata files on CD-ROM or on-line, research centers or licensed sites, and remote access. Rowlands' paper covers a sampling of systems in national statistical offices that permit monitored remote access to confidential microdata. The sample includes six foreign systems and three systems in the United States. The foreign systems are the Luxembourg Income Study, Statistics Canada, Statistics Denmark, Statistics Netherlands, the Australian Bureau of Statistics, and Statistics Sweden. The U.S. agencies are the National

Center for Health Statistics, the National Center for Education Statistics, and the Census Bureau.

The paper reviews the type of methodology used in each of the systems because the methodology influences the kinds of access and results given to users. Rowland reviewed the use of each system and the kinds of research that have benefited from remote access to the extent that such information is available.

Joseph Hotz, the formal discussant for the session, emphasized the tradeoffs associated with different confidentiality protection methods. He said he was struck by the number of dimensions on which the various approaches differ, which makes the process of making an "optimal" decision difficult: there is no simple way of deciding on "the right method." Across these different modes—from public access to data enclaves to remote access to licensing—there are differences not only in terms of degree of access, but also in ease of use, cost, appropriateness for the types of data and information available, ability to customize versus having to rely on standardized data, etc. The alternative methods also differ substantially with regard to how they are financed and how they might be financed. Hotz concluded that in evaluating different methods, one has to consider much more than simply access.

SESSION III:
MEASURING THE RISKS AND COSTS OF DISCLOSURE

The paper by Jerome Reiter, "Estimating Probabilities of Identification for Microdata," describes methods for measuring identification disclosure risks, including those associated with re-identifications from matching to external databases with public-use microdata. The paper describes general methods for calculating sampled units' probabilities of re-identification from the released data, given assumptions about intruder behavior.

When agencies release microdata to the public, intruders may be able to match the information in those data to records in external databases. Reiter presented specific methods for altering data to prevent such matching, including global recoding of variables, data swapping, and adding random noise. He illustrated the methods with data from the Current Population Survey, including random swapping of a subset of the values of variables needed to protect sample "uniques" (across combinations of variables such as age, sex, race, marital status) and using an age recode in addition to swapping to provide the swaps with good protection . He noted that knowing property taxes greatly increases probabilities of re-identification, and adding noise to positive tax values is not sufficient for eliminating uniques, though top coding helps.

William Seltzer and Margo Anderson presented the paper, "Government Statistics and Individual Safety: Revisiting the Historical Record of Disclosure, Harm and Risk," which examines the sparse but important historical record of disclosure, harm, and risk. In the broadest terms, the paper has two interrelated objectives: presentation of a body of facts and presentation of a reconceptualization of a number of the issues related to disclosure and statistical confidentiality in order to understand the implications of the facts assembled. The latter is rooted in the ethical, statistical policy, and statutory origins of the idea of statistical confidentiality.

The focus of the presentation was on issues of disclosure, harm, and risk that have emerged from the use of government statistical agencies or programs to assist in the nonstatistical task of targeting individuals or population subgroups for administrative action. The paper sets out the available evidence concerning such government efforts, which the authors argued have led to serious human rights abuses. Seltzer and Anderson also described a number of barriers to the study of disclosures, harms, and risks associated with government activities.

George Duncan, as the formal discussant, framed his comments on how to evaluate disclosure limitation methods in the context of measuring the risks and costs of disclosure. He cited limitations in current methods for measuring the presence of population uniques: most methods ignore the knowledge state of data snoopers (e.g., a snooper may or may not know that the target individual is in a sample); they provide little information about continuous data; and they provide minimal guidance for evaluating alternative disclosure limitation procedures. Duncan applauded Reiter's application (using data from the March 2000 Current Population Survey) to demonstrate a framework based on probability of identity disclosure and for rigorously exploring the efficiency of such disclosure limitation approaches as global recoding, data swapping, and adding random noise.

SESSION IV:
THE IMPACT OF MULTIPLE IMPUTATION ON DISCLOSURE RISK

The presenter and discussants in this session focused on the advantages and disadvantages of using synthetic data as a method of protecting confidentiality while at the same time providing greater access to data and preserving their informational utility. The presentation and discussion concentrated on three questions: Could use of a multiple imputation method improve data confidentiality without significantly compromising informational utility? How well do the statistical inferences from multiply imputed data match the results that are obtained using the original data? Do multiple imputations provide proper balance between data confidentiality and accessibility?

The paper presented by Trivellore Raghunathan, "Evaluation of Inferences from Multiple Synthetic Data Sets Created Using a Semiparametric Approach," examined evidence on the difference in modeling results with original data and masked data. Techniques of data alteration—such as data swapping, post-randomization, masking, subseparation, truncation, rounding, and collapsing categories—may protect confidentiality, but they may also introduce bias in statistical inferences. The idea of using multiple imputations to create synthetic data sets for public release was introduced by Rubin (1993). The paper reviews pioneering work by Rubin (1993) and Little (1993) developing the methodology, presents extensions, and evaluates the methodology with simulated data sets. Raghunathan outlined the general-purpose semiparametric approach for creating multiple synthetic data sets and showed it to be especially useful when underlying relationships are nonlinear. The goal of Raghunathan's approach was dual: to protect confidentiality and to preserve the key statistical properties of the original data.

Raghunathan mentioned several advantages of creating synthetic samples. For example, one can link the data, synthesize the linked data, and enhance the missing data in the original data file (as is currently being attempted for the HRS and Supplemental Security Insurance (SSI) variables). For that application, the plan is to take some HRS public-use data and the SSI data and then create a full synthesis of that data set.

Although the method of generating synthetic data sets is computationally intensive, Raghunathan emphasized that these multiple data sets can be analyzed using existing software packages with little additional effort. Moreover, he suggested, users of synthetic data should be able to construct an unbiased estimate from the altered data without knowing what exact alteration procedure was used to protect confidentiality.

John Rust, serving as the formal discussant, agreed with Raghunathan's goal of being able to have some statistical procedure that protects confidentiality without altering inference, but he expressed strong distrust of any completely mechanistic procedure to generate synthetic samples. He said that multiple imputation methods might work for some data sets, but he sees many problems with the application of this approach to such complex data as, for example, the HRS.

During the general discussion, Michael Hurd suggested—and several other discussants supported—the idea of an experiment whereby multiple data sets are imputed from actual data, and then one group of researchers analyses the actual data, while another group does the same analysis on the synthetic data. Then the differences between their results could be compared. Discussants agreed that such a test would provide a lot of valuable information. John Abowd said that such experiments are already under way at the Census Bureau, with a link between data from the Survey of Income and Program Participation (SIPP) and detailed Social Security earnings records and other administrative data from the SSA. An extension of the above experiment, proposed by George Duncan, would be to also bring in data snoopers, using whatever tools they might have, to try to identify records in both data sets.

SESSION V:
EMPIRICAL ASSESSMENTS OF THE BENEFITS OF RESEARCH ACCESS TO LONGITUDINAL MICRODATA

John Bailar presented a paper, "The Role of Data Access in Scientific Replication," that describes the underlying issues raised by the role of access to data in scientific replication and, more broadly, the value of scientific replication. His focus was on data generated by nongovernmental sources, primarily in academia, and balancing the concerns of those who generate the data against the public interest in broader use. He noted that the state of understanding about this aspect of academic research is nowhere near as advanced as thinking about confidentiality of and access to federal microdata.

Bailar addressed several conflicts that arise in the context of data access in scientific replication. One such conflict is that society has a strong interest both in protecting privacy and confidentiality and in assuring that scientific findings and interpretations are as close to correct as the state of the art allows. Another conflict is that much research information has personal and proprietary value, which creates barriers to broad access to the data. A third conflict reflects the fact that data are the stock in trade for most research scientists, and scientific rewards are based almost exclusively on the generation and interpretation of data.

Bailar concluded with several propositions. First, few researchers would be happy to give away their final data—and especially the intermediate products of their investigations—if the products of their work are going to be examined by hostile interests bent on destroying the credibility of the findings. Nor is hostile scrutiny likely to advance the state of the science. It may discourage the best scientists from engaging in certain kinds of work that could lead to loss of exclusive access to data. Bailar strongly opposed the view that hostile examination is the best way to uncover the truth. Broad data access also raises questions about being scooped by competitors. Although this is certainly a big concern to researchers and often a barrier to sharing the data, it may have little effect on practice for the simple reason that if nobody generates data, everybody will soon be out of business.

Bailar's second proposition was that the people who are good at generating data are not always the best at analyzing the data. He suggested that there may often be good reasons (though with some limitations) for separation of support for data generation from that for analysis. One limitation is that such separation should be considered case by case. He also noted that those who generate data are not always diligent about completing their own work and making the results public.

Charles Brown presented the session's second paper, "The Value of Longitudinal Data for Public Policy Decisions that Have Been Taken over Time," which assessed the effects of research that uses longitudinal data on public policy. Assessing the effects on policy is more difficult than assessing the effects on academic research: legislators (and other decision makers) rarely cite academic papers and, when they do refer to academic work, it is fair to question whether the research changed the vote (or program decision) or whether the vote (or decision) was based on other considerations, which simply prompted reference to supporting research. However, Brown did attempt to identify policy-related findings based on longitudinal data and to ask whether policy appears to have responded to such findings.

Longitudinal data can make two contributions to research. First, they allow more accurate reporting of transitions between states, durations in a particular state, and changes in variables of interest than is typically possible from a single cross-sectional data collection. Second, longitudinal data allow a researcher to control for otherwise-omitted variation in outcomes among individuals, as long as this variation is constant for given individuals. Both of these contributions are evident in the examples discussed in Brown's paper, drawn from five policy areas: welfare reform, job training, unemployment insurance, preschool programs, and retirement.

Though longitudinal data have played an important role in these policy areas, the contribution of research is constrained by a fundamental and inherent tension—policy research often demands prompt "answers," and longitudinal data take time to be collected. Brown made several suggestions that he said would be particularly helpful for strengthening the contributions of longitudinal microdata to policy analysis:

- Persistence in studying long-run effects. Often, because of funding issues, data are not collected over a long enough period to fully exploit the opportunities to study long-run effects.
- Mining regulatory data. Academic researchers can make impor-

APPENDIX A   107

tant contributions to policy debates about regulatory activities if more data can be made available.
• Matching. Data linking opens up many research opportunities. As an example, creation of data about firms matched to workers' records would be extremely helpful to a range of research questions about business dynamics and the economy.

Dan Newlon, the formal discussant for the session, agreed with John Bailar that researchers should, on publishing their results, make data available at data archives, so that researchers who want to verify the results or extend them can do so. Newlon pointed out that the NSF funded a study by Bill Dewald, then editor of the *Journal of Money, Credit, and Banking*, of the replicability of research results published in the journal. The disturbing surprise was that a third of the authors were unable or unwilling to provide data to support their published results. Another third of authors provided data, but without adequate documentation, so that it was impossible to replicate the published results.

Newlon disagreed with Bailar's position that researchers should not be forced to share their data with others and that the value of giving other researchers access to data was outweighed by the possibility of critical scrutiny that would require investigators to divert energy, time, and effort away from their own research. Newlon explained the essence of the current NSF policy on data sharing: there is a grace period, but once a researcher's grant is finished and the researcher has started publishing results based on those data, then the data are expected to be in the public domain so others can use the data and extend and check the validity of the results.

During the open discussion, Richard Suzman provided another example of when replication in the form of a meta-analysis has been done and is needed—research on the levels of disability in the older population, and trends of disability. Many studies have been done, and they provide very different results. There have been concerns that some survey results could not be replicated: the issue is not just one of making the data available, but also of making the documentation clear.

Keith Rust pointed out that the *Journal of Applied Econometrics* has an online data archive, and the journal just introduced a replication policy as well, which encourages submission of articles replicating results. Suzman supported the idea of withholding some fraction of a grant award that involves data collection (to ensure funds to make the data available), although there are a few data sets that, if they had to be shared, would never be collected in the first place. He also recommended reprinting both "*Sharing Research Data*" (National Research Council, 1985) and "*Private Lives and Public Policies: Confidentiality and Accessibility of Government Sta-*

tistics" (National Research Council, 1993) and using them as a required component in training grants.

Richard Rockwell commented on the costs to data producers of archiving data. In his experience, almost all of these very real costs revolve around the documentation, not the data. For replication, and for secondary analysis of all sorts, researchers and data producers need funding to enable them to archive their data in a usable form.

Trivellore Raghunathan added the point that the current situation with data sharing is much better in social science than in medical science, where the prevailing attitude seems to be that "it is my data, and I have a 25-year plan for the data analysis, and only after the data-analysis plan is exhausted can I think about sharing the data." He said he finds it disturbing that policy decisions can be made on the basis of some data analyses, but researchers and others cannot verify and replicate the findings that underpin those decisions.

SESSION VI:
ASSESSING RESEARCH AND POLICY NEEDS AND CONFIDENTIALITY CONCERNS—
THE ECONOMICS OF DATA ACCESS

The final session of the workshop was designed to facilitate discussion of the tradeoff between societal benefits of data dissemination and confidentiality concerns. It began with the presentation of a paper by Ramon Barquin (coauthored by Clayton Northouse), "Data Collection and Analysis: Balancing Individual Rights and Societal Benefits." Barquin focused on government data collections, which provide the basis for analyzing factors involved in such issues as poverty, health care, education, traffic, public safety, and the environment. He described five benefits of data dissemination:

- The wide dissemination of government statistical data informs policy research.
- The findings that emerge from the analysis of statistical data undergo reexamination and reinvigoration when disseminated to the research community; this process improves data quality by exposing errors.
- Data that are used for one purpose can be put to other uses without substantial investments in new data collection. In addition, data can be combined and result in much more powerful tools for examining the problems facing society.
- When research techniques are shared along with the data, the research community and other data centers improve and hone their own techniques.

- When data are shared with the research community, it will actively involve researchers in the problems confronting the nation and policy makers.

Barquin outlined a framework for how government agencies can attempt to balance the privacy concerns of the individual with the societal good generated from the use of data, offering three guidelines. First, establishing contractual relations with nongovernmental researchers offers a wealth of opportunity without causing undue risks to privacy and confidentiality. The process of applying for and receiving unfettered access to limited sets of government statistical data should force researchers to fully justify their projects and should demonstrate why it is necessary for the researcher to have access to all the data, rather than to a restricted set of the data with identifiers blurred or stripped. The contract should also rigorously uphold the principles of informational privacy, namely, security, accountability, and consent. Second, the Census Bureau's efforts to establish research data centers across the United States offer a fruitful opportunity to share Census Bureau data and provide a good model for the sharing of other types of government statistical data. Third, ultimately, in balancing the public good and individual rights, data collection institutions must effectively manage the three components of this balance: they must supply the technology, provide the correct policy, and cultivate an ethical environment of good will and trust.

Julia Lane presented a paper (coauthored with John Abowd), "The Economics of Data Confidentiality," in which she focused on cost-benefit analysis of data dissemination and confidentiality. In considering how statistical agencies might pursue optimal policies, Lane stated that this goal relies on accurate assessment of the benefits derived from the use of such data, the risks of access (and other costs), and the tradeoff between the two.

Lane noted the substantial social benefits associated with releasing microdata (benefits that are not always realized by the statistical institutes themselves), citing examples similar to those noted by other presenters:

- it permits analysis of complex questions;
- it allow researchers to calculate marginal, not just average effects;
- it creates scientific safeguards when it ensures that other scientists can replicate research findings;
- it promotes improvements to data quality: although statistical institutes expend enormous resources to ensure that they produce the best feasible product, there is no substitute for actual research use of microdata to identify data anomalies;

- it promotes development of a core constituency: the funding of a statistical agency depends on the development of a constituency and greater use of data, which includes the creation of new products from existing data and fosters a broader constituency beyond those who directly have access to the data.

There are three costs of microdata use that must be weighed against the benefits of providing access. First is the cost of providing access. Clearly, the cost of providing access depends on the modality, and several have been developed by statistical institutes across the world: public-use microdata, licensing, remote access systems, and research data centers. The second cost is that of reputation. Most agencies expend enormous effort to make sure that published statistics with their imprimatur are of a high quality and take precautions to protect the confidentiality of the data; they would also have to expend sufficient funds to provide access. The third cost is that of the potential disclosure of respondent identities. The ultimate cost to an agency is for an external researcher to disclose the identity of a business or individual respondent. While the penalties for this are typically substantial—ranging up to 10 years in jail and a $250,000 fine in the United States—the consequences of such a breach could be devastating to respondent trust and response rates.

Lane said it is clear that statistical agencies will increasingly be challenged to provide more access to microdata. This pressure provides a chance to fulfill a critical societal mission. However, since increased access does not come without increased costs, it would seem reasonable to try to control these costs by combining research efforts. Some areas in which joint research and development might provide substantial dividends, for example, include:

- the creation of inference-valid synthetic data sets;
- the protection of microdata that are integrated across several dimensions (such as workers, firms, and geography);
- the quantification of the risk-quality tradeoff in confidentiality protection approaches; and
- the effect on response rates of increased microdata access.

Michael Hurd served as the discussant for the Barquin and Lane presentations. Commenting on Barquin's paper, he expressed the view that, although the issues of ethics and trust are important, self-interest is a powerful mechanism limiting the risk of data disclosure. For an academic researcher, being involved in data disclosure through improper use of data would seriously impair that person's career.

Hurd noted that one omission from both papers is the role of training and education. Researchers ought to be trained in ethics, but they also ought to be continually trained in proper procedures, so that they are aware that this is a serious issue.

On the Abowd and Lane paper Hurd remarked that, at the individual level, researchers want more data, and the reason they want more data is that they get great benefits from more data but bear very little of the cost if something goes wrong. In contrast, agencies are more or less in the opposite position. This creates the tension between the two groups. In economists' terms, those interests need to be internalized in a utility framework in which, as a society, people can make a more informed decision that benefits society rather than the individual actors, namely, the researchers and the agencies. John Rust suggested broadening the two dimensions and including politics, which is another form of self-interest; it provides incentives that determine not just what data are released, but what data are collected. This political dimension can also drive the enterprise to inefficient solutions.

Abowd argued that one of the distinctions that Lane and he are trying to make is to distinguish the different analytic frameworks that economists and statisticians bring to this problem—not because they are in conflict, but because they actually are thinking about two different parts of the problem. Abowd said that statisticians have helped enormously in quantifying the tradeoffs in data production associated with risk measures and the information-loss measures from the basic data. But the economic tradeoffs involve other considerations, such as the benefits to society of research and the costs to individuals of disclosure.

Newlon noted that one needs to distinguish between academic users and other users. In the case of the academic user, there are reputational effects. In the Nordic countries, an academic researcher does not have to have a sworn status to access the data that government employees have because the academic user is viewed as part of the same research and policy advisory community. That is the kind of long-term goal he would like to see emerge.

Suzman raised the sociological issue of what sorts of data people don't want to be released about themselves, and why. There seem to be huge variations in what can and cannot be asked among different people or groups. The ethos of what people consider to be confidential is simply not understood. For example, in some states property values and property taxes are publicly available on the Web, yet these appear to be "confidential" data in other states. That is an area that requires more study.

PAPERS AND PARTICIPANTS

Authors and Papers

John M. Abowd and Julia Lane, "The Economics of Data Confidentiality"
John Bailar, "The Role of Data Access in Scientific Replication"
Ramon Barquin and Clayton Northouse, "Data Collection and Analysis: Balancing Individual Rights and Societal Benefits"
Charles Brown, "The Value of Longitudinal Data for Public Policy Decisions that Have Been Taken over Time"
Andrew Hildreth, "The Census Research Data Center Network: Problems, Possibilities and Precedents"
David McMillen, "Privacy, Confidentiality, and Data Sharing"
Henry H. Perritt, Jr., "Efficacy of Different Theories of Enforcement"
Trivellore Raghunathan, "Evaluation of Inferences from Multiple Synthetic Data Sets Created Using a Semiparametric Approach"
Jerome Reiter, "Estimating Probabilities of Identification for Microdata"
Sandra Rowland, "An Examination of Monitored, Remote Microdata Access Systems"
Marilyn Seastrom, Candice Wright, and John Melnicki, "The Role of Licensing and Enforcement Mechanisms in Promoting Access and Protecting Confidentiality"
William Seltzer and Margo Anderson, "Government Statistics and Individual Safety: Revisiting the Historical Record of Disclosure, Harm and Risk"

Participants

John M. Abowd (panel member) is Edmund Ezra Day professor of industrial and labor relations at Cornell University, director of the Cornell Institute for Social and Economic Research (CISER), and a distinguished senior research fellow at the U.S. Census Bureau.

Margo Anderson is a professor of history and director of the Urban Studies Program at the University of Wisconsin-Milwaukee.

John Bailar is professor emeritus at the University of Chicago. His research is in the fields of medicine and statistics, and he has written extensively about science conduct and ethics.

Ramon Barquin is president of Barquin International and was previously the first president of the Data Warehouse Institute. His work is directed to developing information system strategies and data warehousing for the public and private sectors.

Charles Brown is a professor of economics and program director at

APPENDIX A

the Survey Research Center at the University of Michigan whose research focuses on a wide range of topics in empirical labor economics.

Joe S. Cecil (panel member) is project director in the Program on Scientific and Technical Evidence of the Division of Research at the Federal Judicial Center. He is responsible for judicial education and training in the area of scientific and technical evidence.

George T. Duncan (panel member) is a professor of statistics in the H. John Heinz III School of Public Policy and Management and the Department of Statistics at Carnegie Mellon University.

Andrew Hildreth is research director at the California Census Research Data Center and a professor in the Department of Economics at the University of California at Berkeley.

V. Joseph Hotz (panel member) is a professor and chair of the Department of Economics at the University of California at Los Angeles.

Michael Hurd (panel member) is senior economist and director for the RAND Center for the Study of Aging.

J. Bradford Jensen is deputy director of the Institute for International Economics, having recently moved there from serving as director of the Center for Economic Studies at the U.S. Census Bureau. At the Census Bureau, he directed the center's internal and external research programs, managed its Research Data Center network, and supervised its relationships with collaborating universities and research organizations.

Diane Lambert (panel member) is the director of statistics and data mining research at Bell Labs. She has made seminal contributions to fundamental statistics theory and methods and has been a leader in defining a role for statistics in data mining and massive data problems.

Julia Lane is a principal research associate in the Labor and Social Policy Center at the Urban Institute, concentrating in the areas of income and wealth distribution, labor markets, employment, and education.

Michael Larsen is a professor in the Department of Statistics at Iowa State, working in the areas of survey sampling, administrative records and record linkage, missing data problems, finite mixture models and latent class models, small-area estimation, and Bayesian statistical modeling.

David McMillen is the information and government organization specialist with the Committee on Government Reform and Oversight, U.S. House of Representatives, and the government information specialist for Henry A. Waxman (D-CA). He covers issues involving the collection, dissemination, and preservation of government information, and he has worked extensively on legislation relating to confidentiality and data sharing.

John Melnicki is president and CEO of Harbor Lane Associates, Inc., in Washington, D.C. In addition, he is the senior security advisor for restricted data for the U.S. Department of Education, the National Science Foundation, and various research and educational institutions around the world.

Dan Newlon is program director for economic science at the U.S. National Science Foundation, where his job is to select directions for investment in research.

Henry H. Perritt, Jr., is a professor of law and vice provost at the Illinois Institute of Technology and director of the Center for Law and Financial Markets at Chicago-Kent College of Law.

Jerome Reiter is a professor at the Institute of Statistics and Decision Sciences at Duke University. He has worked with the U.S. Census Bureau and recently joined the Digital Government Project of the National Institute of Statistical Sciences.

Trivellore Raghunathan is a research professor at the Institute for Social Research and professor of biostatistics, both at the University of Michigan.

Richard C. Rockwell (panel member) is a professor of sociology at the University of Connecticut.

Sandra Rowland recently retired from the U.S. Census Bureau, where she was the Internet data dissemination system team leader. Among other duties, she managed the advance query interactive web system.

John Rust is a professor of economics at the University of Maryland, with major research interests in numerical dynamic programming and retirement behavior.

Marilyn Seastrom is chief statistician and program director for the Statistical Standards Program at the National Center for Education Statistics, U.S. Department of Education. She has written extensively on data access, licensing, and confidentiality issues.

William Seltzer is a senior research scholar at the Institute for Social Research of the Department of Sociology and Anthropology at Fordham University.

Eleanor Singer (panel chair) is a research professor at the Survey Research Center of the Institute for Social Research at the University of Michigan. Her research focuses on motivation for survey participation and has touched on many of the important issues in survey methodology, such as informed consent, incentives, interviewer effects, and nonresponse bias.

Richard Suzman is associate director for Behavioral and Social Research at the National Institute on Aging (NIA). Suzman developed NIA's Economics of Aging program, one of the first of its kind to look at socioeconomic factors and health.

Katherine Wallman is chief statistician at the U.S. Office of Management and Budget. She is responsible for overseeing and coordinating federal statistical policies, standards, and programs; developing and fostering long-term improvements in federal statistical activities; and representing the federal government in international organizations.

Candice Wright is an analyst at the U.S. Office of Management and Budget. She recently completed her M.S. in public policy from Carnegie Mellon University and holds a B.S. in management from Bentley College. Her current interests include data privacy and information security.

APPENDIX B

Biographical Sketches of Panel Members and Staff

Eleanor Singer (*Chair*) is a research professor at the Survey Research Center of the Institute for Social Research at the University of Michigan. Her research focuses on motivation for survey participation and has touched on many of the important issues in survey methodology, such as informed consent, incentives, interviewer effects, and nonresponse bias. Two of her major studies examined the role of privacy and confidentiality concerns as factors in response to the 1990 and 2000 decennial censuses, and she was a member of the National Academies panel that produced *Private Lives and Public Policies: Confidentiality and Accessibility of Government Statistics*. She is most recently a coauthor of *Survey Methodology* (with Robert M. Groves and others) and a coeditor of *Methods for Testing and Evaluating Survey Questionnaires* (with Stanley Presser and others). She is a past president of the American Association for Public Opinion Research and a recipient of its award for exceptionally distinguished achievement. She holds a B.A. degree from Queens College and a Ph.D. degree in sociology from Columbia University.

John M. Abowd is the Edmund Ezra Day professor of industrial and labor relations at Cornell University and director of the university's Institute for Social and Economic Research. He is also a distinguished senior research fellow at the U.S. Census Bureau, a research associate at the National Bureau of Economic Research in Cambridge, MA, and a research affiliate at the Centre de Recherche en Economie et Statistique in Paris, France. Previously, he was also on the faculty of Cornell's Johnson Graduate School of Management. Professor Abowd's current research focuses on the creation and use of linked, longitudinal data on employees and employers. His other research interests include international comparisons of labor market outcomes; executive compensation, again with a focus on international comparisons; bargaining and other wage-setting institutions; and the econometric tools of labor market analysis.

Joe C. Cecil is project director of the Program on Scientific and Technical Evidence in the Division of Research of the Federal Judicial Center (FJC), in Washington, DC. In that position he is responsible for judicial education and training about scientific and technical evidence and the lead staff for the FJC's Reference Manual on Scientific Evidence, which is the primary source book on scientific evidence for federal judges. He is the author of numerous publications concerning legal standards affecting exchange of information for research purposes. Other areas of interest include the use of scientific and technical evidence in litigation, variations in procedures used by federal courts of appeals, and management of mass tort litigation. He holds a Ph.D. degree in psychology and a J.D. degree, both from Northwestern University.

Constance F. Citro *(Staff Director)* is director of the Committee on National Statistics. She is a former vice president and deputy director of Mathematica Policy Research, Inc., and was an American Statistical Association/National Science Foundation research fellow at the U.S. Census Bureau. For the committee, she has served as study director for numerous projects, including the Panel to Review the 2000 Census, the Panel on Estimates of Poverty for Small Geographic Areas, the Panel on Poverty and Family Assistance, the Panel to Evaluate the Survey of Income and Program Participation, the Panel to Evaluate Microsimulation Models for Social Welfare Programs, and the Panel on Decennial Census Methodology. Her research has focused on the quality and accessibility of large, complex microdata files, as well as analysis related to income and poverty measurement. She is a fellow of the American Statistical Association. She has a B.A. degree from the University of Rochester and M.A. and Ph.D. degrees in political science from Yale University.

George T. Duncan is a professor of statistics in the H. John Heinz III School of Public Policy and Management and the Department of Statistics at Carnegie Mellon University. His current research work centers on information technology and social accountability. He has lectured in Brazil, Italy, Turkey, Ireland, Mexico and Japan, among other places. Duncan chaired the Panel on Confidentiality and Data Access of the National Academies (1989-1993), which produced *Private Lives and Public Policies: Confidentiality and Accessibility of Government Statistics,* and he chaired the

American Statistical Association's Committee on Privacy and Confidentiality. He is a fellow of the American Statistical Association, an elected member of the International Statistical Institute, and a Fellow of the American Association for the Advancement of Science. In 1996 he was elected *Pittsburgh Statistician of the Year* by the American Statistical Association. He received B.S. and M.S. degrees from the University of Chicago and a Ph.D. degree from the University of Minnesota, all in statistics.

Eugenia Grohman *(Study Director)* is associate executive director of the Division of Behavioral and Social Sciences and Education at the National Research Council. She served as study director for the panel during the last stages of its work. She has worked on many previous reports of the Committee on National Statistics, including *Sharing Research Data* and *Private Lives and Public Policies: Confidentiality and Accessibility of Government Statistics*. She attended the University of Chicago and received a B.A. degree in political science from the University of California at Los Angeles.

V. Joseph Hotz is a professor in the Department of Economics at the University of California at Los Angeles. He also serves as a principal investigator of the California Census Research Data Center. His work concentrates on the economics of the family, applied econometrics, and the evaluation of social programs. His extensive published work has examined the relationship between the labor force participation and childbearing patterns of married women; the effect of working while in school on the subsequent wages of men in the United States; and methods for assessing the causal effects of social programs. His most recent work has focused on assessing the effects of child care regulations on children's accident rates; the effects of welfare-to-work programs on the labor market successes of past welfare recipients; the strategic interactions of parents and adolescents over the latters' engagement in risky behavior; and evaluation of the employment effects of the Earned Income Tax Credit (EITC) Program.

Michael Hurd is senior economist and director for the Center for the Study of Aging of RAND and a research associate at the National Bureau of Economic Research. Previously, he was professor of economics at the State University of New York at Stony Brook. He is a member of the National Academy of Social Insurance, the steering committee for the Health and Retirement Study, and he was a member of the Technical Panel for the Social Security Advisory Council in 1990-1991. He has served as consultant to the National Institute on Aging on re-interviewing in the Retirement History Survey and to the Social Security Administration on re-interviewing in the New Beneficiary Survey. His research involves income and wealth of the elderly and pensions and retirement economics. He received a Ph.D. degree in economics from the University of California at Berkeley.

Diane Lambert is the director of statistics and data mining research at Bell Labs. She has made seminal contributions to fundamental statistics theory and methods and has been a leader in defining a role for statistics in data mining and massive data problems. She continues to introduce significant technological innovations in statistics, as well as fostering a close relationship between research and various business units. Lambert holds five patents. She previously served as editor of the *Journal of the American Statistical Association,* and she is a fellow of both the American Statistical Association and the Institute of Mathematical Sciences. She holds a Ph.D. degree from the University of Rochester in New York.

Christopher Mackie (*Study Director*) is on the staff of the Committee on National Statistics and served as study director for most of the panel's life. He served as study director for a number of economic measurement projects, including those that produced the reports, *At What Price? Conceptualizing and Measuring Cost-of-Living and Price Indexes,* and *Beyond the Market: Designing Nonmarket Accounts for the United States.* Prior to joining CNSTAT, he was a senior economist with SAG Corporation, where he conducted a variety of econometric studies in the areas of labor and personnel economics, primarily for federal agencies. He is the author of *Canonizing Economic Theory.* He has a Ph.D. in economics from the University of North Carolina and has held teaching positions at the University of North Carolina, North Carolina State University, and Tulane University.

Kenneth Prewitt is the Carnegie professor of public affairs at the School of International and Public Affairs at Columbia University. He was director of the U.S. Census Bureau from 1998 to January 2001. His government service followed a career in higher education and private philanthropy, including: president of the Social Science Research Council, senior vice president of the Rockefeller Foundation, director of the National Opinion Research Center, based at the University of Chicago, and professorships at the University of Chicago, Stanford University, Washington University, the University of Nairobi, and Makerere University in Uganda. His current research is on the policy consequences of racial classification in official statistics, and he recently published *Science and Politics in Census-Taking.* He holds a Ph.D. degree from Stanford University.

Richard Rockwell is a professor of sociology at the University of Connecticut. Previously, he was executive director of the Institute for Social

Inquiry/Roper Center for Public Opinion Research and director of the Inter-University Consortium for Political and Social Research at the University of Michigan. One of the nation's foremost experts on social science and public opinion research, he has published numerous articles on social science methodology and has designed related software programs. He holds a Ph.D. degree in sociology from the University of Texas at Austin.